Kiat Seng Yeo

Millimeter-Wave IC Design Techniques for Beam-Forming Applications

Fanyi Meng
Kaixue Ma
Kiat Seng Yeo

Millimeter-Wave IC Design Techniques for Beam-Forming Applications

LAP LAMBERT Academic Publishing

Impressum / Imprint
Bibliografische Information der Deutschen Nationalbibliothek: Die Deutsche Nationalbibliothek verzeichnet diese Publikation in der Deutschen Nationalbibliografie; detaillierte bibliografische Daten sind im Internet über http://dnb.d-nb.de abrufbar.

Bibliographic information published by the Deutsche Nationalbibliothek: The Deutsche Nationalbibliothek lists this publication in the Deutsche Nationalbibliografie; detailed bibliographic data are available in the Internet at http://dnb.d-nb.de.

Coverbild / Cover image: www.ingimage.com

Verlag / Publisher:
LAP LAMBERT Academic Publishing
ist ein Imprint der / is a trademark of
OmniScriptum GmbH & Co. KG
Bahnhofstraße 28, 66111 Saarbrücken, Deutschland / Germany
Email: info@omniscriptum.com

Herstellung: siehe letzte Seite /
Printed at: see last page
ISBN: 978-3-330-02235-5

Zugl. / Approved by: Singapore, Nanyang Technological University, 2015

TABLE OF CONTENTS

CHAPTER 1

Introduction

1.1 Millimeter-wave/Terahertz Applications

The radio-frequency integrated circuit (RFIC) designs have achieved a remarkable advancement at a tremendous speed. It enables realization of low-lost commercial portable wireless communication system, as well as advanced wireless communications for military application, working environment, and security monitoring. As data transferring becomes more bulky and complex, the demand to support high frequency high bandwidth traffic requires a magnificent enhancement of system performance at an exponential rate.

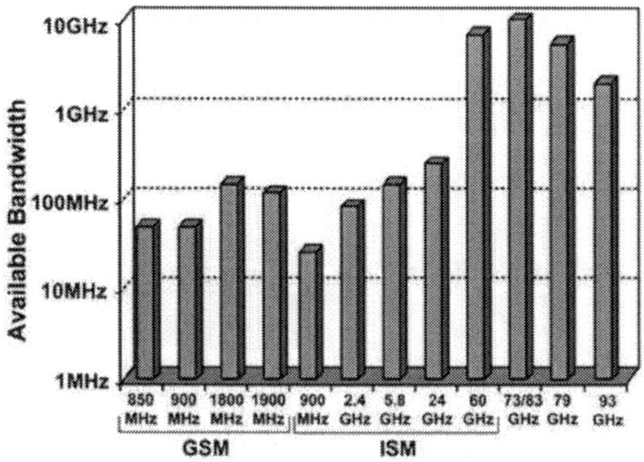

Figure 1.1 Available bandwidth at different bands.

Figure 1.1 shows the bandwidth at different frequency bands. At millimeter-wave (mm-wave) regime, where the available bandwidth is 10-100 times wider than GSM counterparts, it enables high speed wireless links. Besides, at higher frequency, the size of passive components and packaging that forms an important part of most wireless systems, could be designed much more compact and achieve small form-factor for communication chipsets. Figure 1.2 illustrates the simplified packaging dimensions.

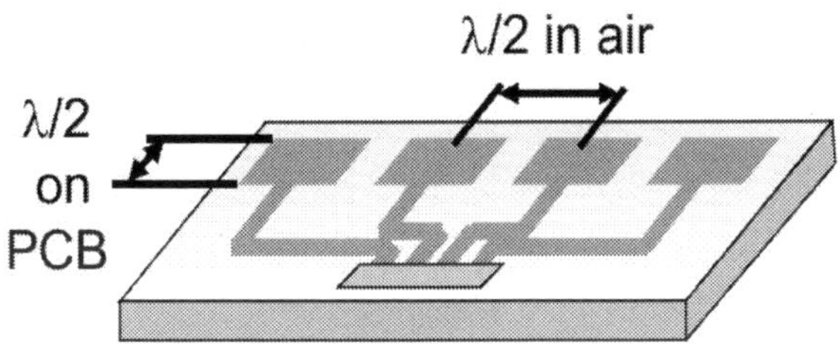

Figure 1.2 Simplified packaging dimension. (λ is the wavelength)

In imaging systems, the resolution is associated with wavelength and bandwidth as indicated in (1.1) and (1.2) as follows:

$$\text{Pixel Resolution} \propto 1/\lambda \tag{1.1}$$

$$\text{Range Resolution} \propto \text{Bandwidth} \tag{1.2}$$

Thus, at mm-wave and terahertz frequencies, both pixel and range resolutions could be enhanced.

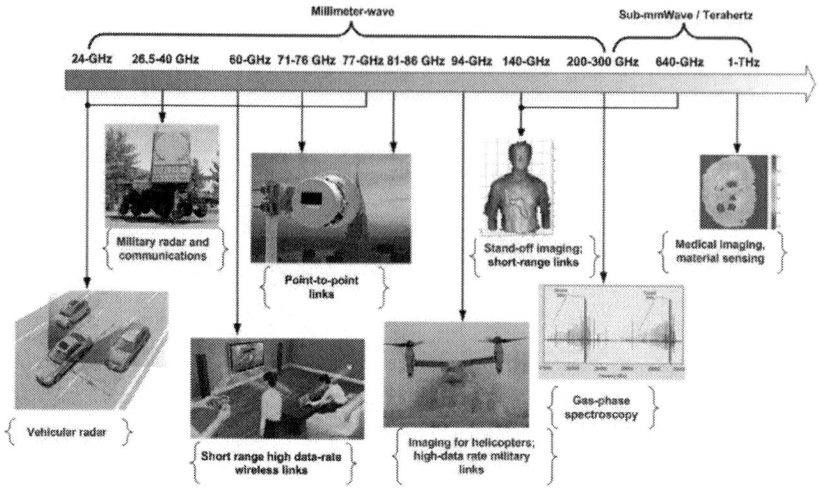

Figure 1.3 Millimeter-wave/Terahertz applications.

Figure 1.3 shows the various application scenarios at mm-wave and terahertz regimes. Due to the availability of wide bandwidth and short wavelength than RF frequencies, the mm-wave circuits and systems excel in the high-data-rate P2P communications, medical imaging, spectroscopy, and radar applications.

One of the most promising systems is the 60-GHz transceivers, utilizing the 57–64 GHz Industrial, Scientific and Medical (ISM) band [1-4]. This 7 GHz wide bandwidth is very attractive for achieving high data-rate wireless communications. However, due to higher path loss (68 dB/m) and atmospheric loss, the 60 GHz band is mostly suitable for short-range applications (~10 m) and point-to-point communication link is necessary unless a phased-array transceiver topology is considered. Figure 1.4 presents the primary 60-GHz applications, such as transferring uncompressed high-definition (HD) video from a set-top box to a display screen and downloading media contents from a kiosk to a portable hand-held device. Note that these applications require a relatively controlled environment (i.e. within a room in a house or an office space), and therefore there are no stringent interference specifications like in cell-phone bands.

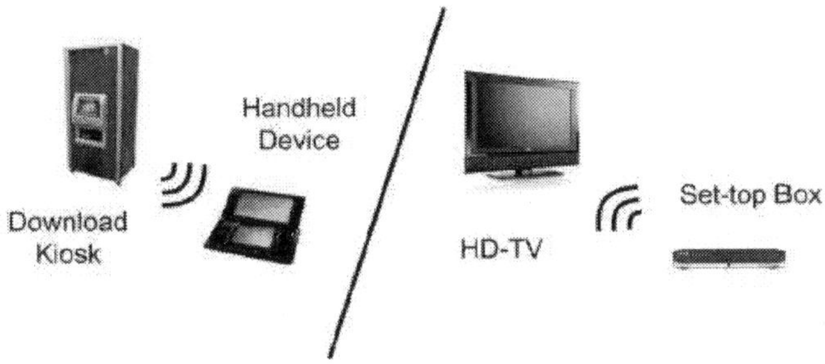

Figure 1.4 Primary applications of 60-GHz communication systems.

1.2 MM-Wave Beam-Forming Techniques

Beam-forming systems that are used for electrical beam steering to various directions within the predefined field view, have become the major solution for millimeter-wave (mm-wave) systems, especially those operating at the unlicensed 60-GHz. Thus systems typically utilize phased-array technique at mm-wave frequencies, and emulate the high-gain antenna with the advantages of array gain, high directivity, and spatial coverage, as shown in Figure 1.5 and Figure 1.6.

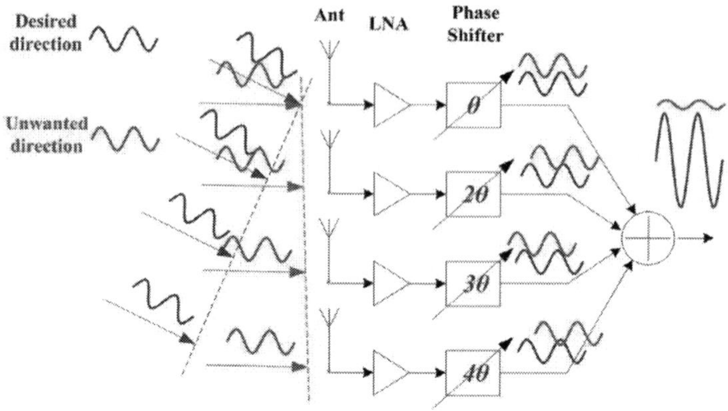

Figure 1.5 High directivity property of beam-forming technique.

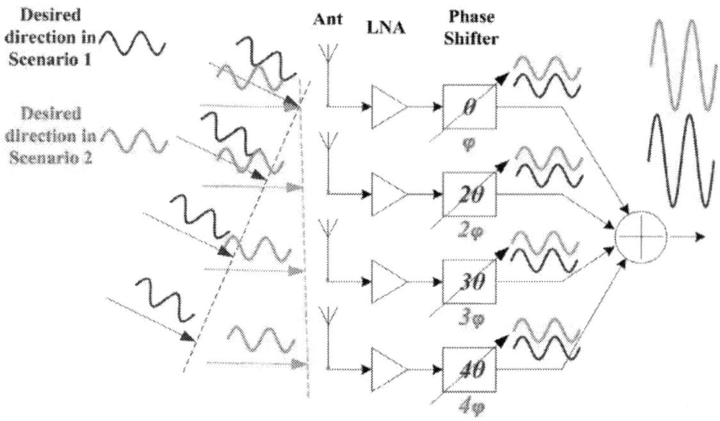

Figure 1.6 Wide coverage property of beam-forming technique.

In addition, phased-array allows non-line-of-sight communication which cannot be established in conventional mm-wave transceivers due to high barrier loss at 60-GHz. Theoretically the performance of transceiver improves as the number of elements (N) in phased-arrays, as follows [5]:

$$\text{RX SNR improvement} = 10 \times \log(N) \quad (1.3)$$

$$\text{TX Array Gain} = 20 \times \log(N) \quad (1.4)$$

Fully integration of phased-array systems with digital baseband in silicon based technologies, leads to numerous improvements in cost, size, power consumption, and reliability. Meanwhile, it provides feasibility to perform fully on-chip signal processing and controlling, without any off-chip components, resulting in additional savings in cost and power. The multiple channels, operating in harmony, provide benefits at both system and circuit level. For example, a 60-GHz integrated phased-array will make gigabit-per-second directional point-to-point communication feasible. At the circuit level, the division of the signal into multiple parallel paths relaxes the signal handling requirements of individual blocks, especially power amplifiers.

Despite its high functionality, phased-array is still a costly solution. Multiple elements arise total power consumption, circuit size, complexity of RF and IF circuits routing and inter-element isolation, and system calibration algorithm. Thus, great challenges stills exits for phased-array to be used in the commercial and mobile applications where the power and cost are primarily concerned.

Lossy silicon substrate and poor power handling capability have to be considered wisely in the circuit designs. Besides, other than the traditional building blocks like low noise amplifier [6], power amplifier [7], frequency synthesizer [8, 9], modulator [2], on-chip filter [10], switches [11], and etc. in single element transceiver [1], new building blocks like the variable phase shifters [12], RF variable gain amplifiers [13], and power combining/splitting circuits [14] have to be designed and co-designed on-chip to provide beam steering capability. These designs must also feature compact size and low power consumption characteristics. Such systems have been demonstrated in two-, four-, sixteen-, and thirty-two-elements phased-arrays [3, 5, 15-24].

In the early works of mm-wave phased-arrays, four-elements receivers and transmitter are successfully implemented in 0.12-μm SiGe BiCMOS [23] and 65-nm CMOS [25], respectively. In [23], the receiver array uses passive RF phase shifting topology, achieving full spatial coverage with peak-to-null ratio higher than 25dB using with power consumption of ~65mW per element. In the transmitter array [25], active phase shifting was adopted due to lower gain per stage in CMOS. The array achieves fully-differential signal processing with independent tuning of both vertical and horizontal polarizations with ~20dB gain, 11dBm output Psat, and power consumption of ~150mW per element.

A fully-integrated sixteen-element phased-array receiver was reported in [5]. The arrays use aperture-coupled patch-antennas packaged with RX IC in multi-layer organic and LTCC, demonstrating EVM better than -18dB in both line-of-sight and non-line-of-sight (using write board as reflector) links of ~7.8m and ~9m spacing, respectively. In [24], a thirty-two-element phased-array TX/RX chip with a 2-bit phase shifter and IF converter to/from 12GHz was implemented. The array achieves 12.5dB gain, 11dB noise figure, -17dBm RX IP1dB, and 8dBm TX array Psat, with total power consumption of ~500mW.

The most recent works of 60-GHz phased-array includes very low power and compact designs that target for short-range (<1m) and portable devices applications [15, 16], and fully-integrated TX/RX chipset with state-of-the-art EVM performance (better than -20dB) in [3].

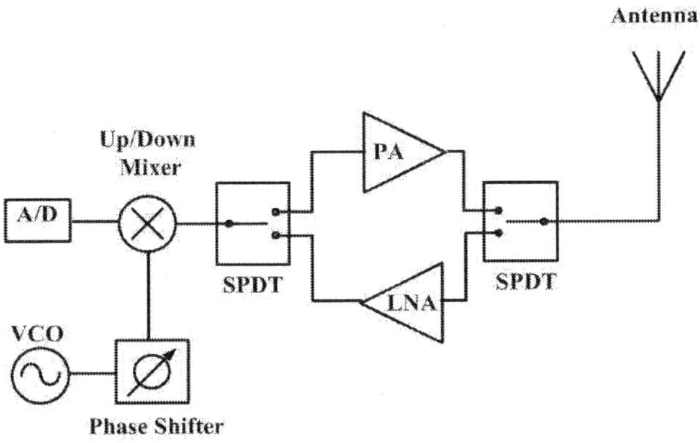

Figure 1.7 A simplified THz communication/detection system.

1.3 Terahertz Systems

Integrated circuits that operate beyond 100 GHz are also called sub-terahertz ICs. Thus systems or even terahertz (0.3-10 THz) ones have up till now not been successfully deployed in the commercial markets. However, the properties of this frequency range have provided great potential applications for high-data-rate short-range communication systems, damage free security detections, and military radar systems. A typical terahertz system is depicted in Figure 1.7.

The main function blocks of the architecture include power amplifiers (PA), low noise amplifiers (LNA), mixers and voltage controlled oscillators (VCO). Among these components, the VCO, phase shifters and the analog to digital/digital to analog convertor are normally operating at lower frequencies, i.e. intermediate frequencies (IF), to minimize the loss of the RF signal. The up/down mixers are essential components of a high frequency system. They transfer the RF frequency signals to the IF band for a receiver, and IF modulation signal to RF for the transmitter. In most cases, a frequency multiplier is included to generate a carrier signal for the system. The single-pole double-throw (SPDT) switches exist in a transceiver to select the transmitting/receiving channel. Passive circuits are commonly used for designing the switches. PA and LNA are the key components for determining the performance of the system. The design of PA and LNA are, however, most challenging for such system design. The performance of the amplifiers is limited by the fabrication technology, namely the unity-gain frequency f_T and the maximum oscillating frequency f_{MAX} of the transistors used in the design. It is extremely difficult to design a PA or LNA beyond 200 GHz and hence, it is a norm to exclude both in transceiver systems [26-29]. This then put strain on the performance of other signal sources. Finally, at the very front end of the transceiver, either an on-chip antenna is integrated or an off-chip horn antenna is used to ensure higher directivity after packaging [29, 30]. The system illustrated above can be used for communication or radar detection.

1.4 CMOS Technologies Leading to MM-wave/THz

As discussed previously, the performance of the MMICs designed is actually limited by the characteristics of the RF transistors used in the circuit blocks. Various technologies provide the process design kit (PDK) for RF/MMIC designers. Silicon-based technologies and III-V compounds are the two that are most commonly used.

The complementary metal oxide silicon (CMOS) process was invented in 1960s and commonly used in RFIC fabrication in recent years. Its greatest potential is its ability for high integration of devices, so most publicity surrounding one-chip solutions for

Bluetooth, ZigBee and other applications with requirements for very low cost, small size and the incorporation of significant digital circuitry, where CMOS excels [31]. As CMOS process continues to scale down, the high f_T and f_{MAX} of transistors provide the potential of high frequency RFIC design. For example, 60GHz systems have been proposed, designed, and fabricated using the CMOS process [1, 3, 15, 17, 25, 32, 33]. However, after following Moore's law for 40 years, CMOS scaling is facing its bottle neck. Although many techniques are being researched to break the limits of CMOS scaling, they are still too costly. The 20-nm CMOS process is even more expensive than the InP process which can meet the same device properties.

1.5 Motivation and Objectives

Tremendous efforts on MM-wave and terahertz IC designs have been aroused both from the academic and the industrial areas. Researchers all over the world are aiming for higher operation frequencies and associated promising applications. The properties of mm-wave and terahertz frequencies guarantee potential applications in high-data-rate short-range communications and imaging systems. The amazes and challenges of such circuits actually draw the author's passions.

However, due to limited transistor gain and bandwidth, it is always difficult to achieve large output power at mm-wave/THz frequencies. As discussed above, the beam-forming technique emulates a high-gain antenna and presents as a good candidate. In beam-forming or phased-array systems, the system performance is directly related to the number of multiple channels, which leads to large amount of silicon area and power consumption. Current researchers have demonstrated some novel compact and low-power building blocks; however, there is still a large gap to be successfully deployed for commercial applications.

With the motivation discussed above, the goal of this dissertation is to propose and design mm-wave and THz blocks with miniaturized circuit areas for applications of high-speed wireless communication. The phase shifters, switches, and amplifiers are the key circuit blocks to be discussed; where the working frequency range is from 60 GHz to 280 GHz. GLOBALFOUNDRIES (GF) commercial 65-nm CMOS technology is preferred to fabricate the designs due to the merits discussed before, which also provides the potential to integrate the whole system on one chip in future.

1.6 Organization of the Book

The remaining chapters of the dissertation are organized as follows:

Chapter 2 presents three sub-terahertz single-pole double-throw (SPDT) switches in 65-nm CMOS. A new topology, namely the magnetically switchable artificial resonator is introduced to alleviate the bulky and lossy transmission lines used in conventional SPDT designs. Besides its compact size, the coupled-line based artificial resonator features lower loss compared to the transmission lines. The small-signal equivalent circuit models of the shunt nFET switches and the resonator were developed and investigated. Further, the auxiliary coupled-lines are introduced to control the magnetic coupling strength between main coupled-lines of the artificial resonator, and improve the switch isolation performance. To validate the theoretical analysis, three switch prototypes are designed and fabricated in a 65-nm CMOS technology; two are at 130-180 GHz and one is at 220-285 GHz. It is verified experimentally that the isolation is improved by ~2 dB with almost zero compensation on switch insertion loss, by using auxiliary coupled-lines in the 130-180 GHz switch designs. The fabricated 130-180 GHz SPDT switch achieves insertion loss of 3.3 dB, and isolation of 23.7 dB. To the authors' best knowledge, the circuit size of 0.0035 mm^2 is the smallest among SPDT switches with similar operating frequencies. The fabricated 220-285 GHz SPDT switch features measured insertion loss of 4.2 dB including RF pad losses, isolation of 19 dB, return loss of better than 10 dB, simulated P1dB of 9.2 dBm, and zero power consumption. To the best of authors' knowledge, this switch achieves the highest operating frequency and smallest chip size among reported SPDT switches in CMOS and BiCMOS technologies.

Chapter 3 presents the design of a compact 60-GHz phase shifter that provides 5-bit digital phase control and 360° phase range for beam-forming systems. The phase shifter is designed by using the proposed cross-coupled bridged T-type topology and switched-varactor reflective-type topology. The topologies are analyzed using small-signal equivalent circuit model. Further, the design equations are derived and investigated. To validate the theoretical analysis, a 60-GHz 5-bit 360° phase shifters are designed in a commercial 65-nm CMOS technology. The fabricated 360° phase shifter features good performance of 32 phase states from 57 to 64 GHz with RMS phase error of 4.4°, total insertion loss of 14.3±2 dB, RMS gain error of 0.5 dB, P_{1dB} of better than 9.5 dBm, and power consumption of almost zero. The average insertion loss is only 2.8 dB per control bit. To authors' best knowledge, the designed 360° phase shifter with the size of 0.094 mm^2 is the smallest 5-bit passive phase shifter at frequencies around 60 GHz.

Chapter 4 reports a 56-67 GHz bidirectional low-noise amplifier power amplifier (LNAPA) design. To eliminate the use of Tx/Rx switches and support dual-mode operation, the bidirectional matching networks are introduced to connect LNA and PA core circuits in parallel and satisfy isolation requirements with full consideration of input/output impedance matching of the LNA and PA. Thus, the operation modes are simply selected by alternated gate biasing of the LNA and PA core circuits. Fabricated in a commercial 65-nm CMOS technology, the Rx mode features peak gain of 21.5 dB with gain of >17 dB over 56-67 GHz, NF of 6.7 dB with 39.6 mW power consumption, while Tx mode achieves peak gain of 24.5 dB with gain of >17 dB over 56-65 GHz, P_{SAT} of 8.4 dBm, PAE of 8.7% with 71.1 mW power consumption. The reverse isolation in both modes is better than 43 dB. The circuit occupies a compact size of 0.22 mm^2.

Finally, Chapter 5 draws the conclusion and gives the area that merits future works.

CHAPTER 2

Sub-Terahertz SPDT Switches

2.1 Introduction

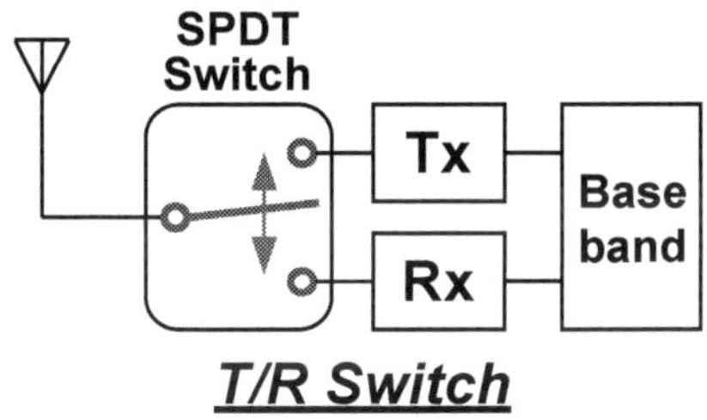

(a)

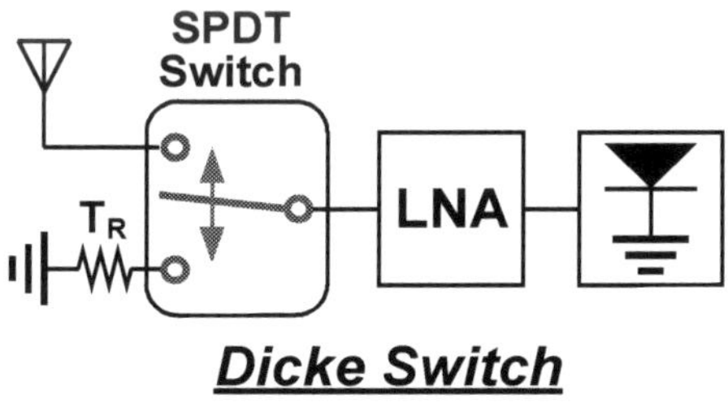

(b)

Figure 2.1 SPDT switches used as (a) T/R switch and (b) Dicke switch.

Single-pole double-throw (SPDT) switches are a key building block in millimeter-wave (mm-wave) transceiver and imager SoCs and SiPs [3, 11, 29, 34-66]. As shown in Figure 2.1, a SPDT switch can be used as a T/R switch to enable transceiver time-division duplex (TDD) operation [34]. In an imaging receiver, it can be used as a Dicke switch to switch between antenna and reference load to eliminate imager fluctuations [29, 35]. To provide acceptable compromises of noise figure, output power and sensitivity in transceivers or imagers, the switches are required to feature an insertion loss of ~3 dB and an isolation of ~20 dB [35].

In the literature, mm-wave SPDT switches [29, 34-38] mostly adopt $\lambda_g/4$ transmission-line (T-line) topology as shown in Figure 2.2, where λ_g is the guided wavelength at operating frequency. When control voltage V_C is set high, the top transistor is turned on which creates low impedance at point *A*. The $\lambda_g/4$ T-line transforms this low impedance into high impedance at common point *T*, which prevents signal flowing to the top path. Meanwhile, the bottom path becomes a low-loss path with the bottom transistor turned off. Shunt T-lines T_M are added to resonate out the parasitic capacitance of transistors. It operates in a similar fashion when V_C is set low. However, this topology heavily relies on T-lines that present relatively bulky circuit size and high loss at mm-wave frequencies [29, 34-38].

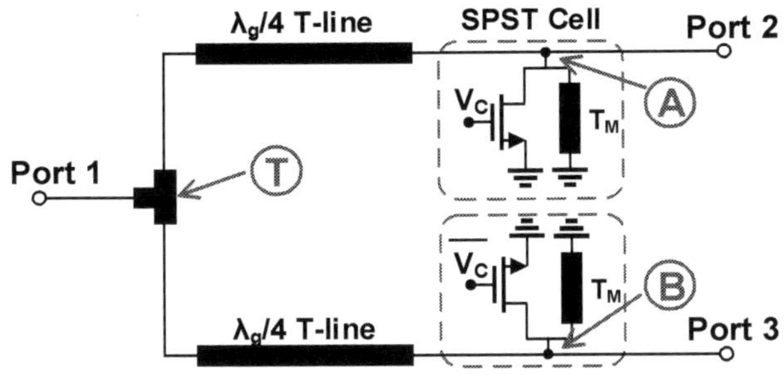

Figure 2.2 Conventional $\lambda_g/4$ T-line topology for mm-wave SPDT switches.

In the past, CMOS SPDT switches were demonstrated at 50-67 GHz [35] and 75-110 GHz [36], with low insertion loss of 1.5 dB and 2.8 dB respectively. However, it is expected that the insertion loss of CMOS switches will be well above 3 dB as the operating frequency goes beyond 110 GHz [34, 39, 40]. Transformer-based switch was explored in [41] where excellent size reduction was achieved. However, the closely spaced transformer ports limit the isolation performance to 13.7 dB only. In another works [40], the $\lambda_g/4$ transmission lines are emulated with lumped inductors and capacitors, where circuit sizes are reduced as a cost of reduced operating bandwidth and increased insertion loss. Recently, advanced processes like HBT and SOI were investigated for switch designs [36-38]. In [36], a 73-110 GHz SPDT switch with insertion loss of only 1.1 dB was presented using 90-nm HBT with reverse-saturated configurations. Several HBT and SOI designs beyond 110 GHz were also reported in [37] and [38], achieving insertion loss of around 3 dB.

In this chapter, a new SPDT topology using a magnetically switchable artificial resonator is reported [42, 43]. The chapter is organized as follows. Section 2.2 presents the theoretical analysis of proposed topology. The proposed resonant structure alleviates the problematic $\lambda_g/4$ and matching T-lines. Magnetic switching concept is adopted in the SPDT design to further improve the switch isolation at almost zero compensation of insertion loss. Section 2.3 presents the systematic design, implementation, performance of a 130-180 GHz prototype in a 65-nm CMOS. To cater for D-band imaging applications, a 220-285 GHz switch is designed and verified experimentally in Section 2.4. Section 3.5 gives a brief summary.

2.2 Topology and Analysis

In Figure 2.3(a), the proposed SPDT switch adopts artificial resonator concept which comprises three coupled-lines, two switch transistors $M_{1\text{-}2}$, one lump capacitor C_C, and bias resistors R_G. Auxiliary coupled-lines which only affect the magnetic coupling between the three main coupled-lines in Figure 2.3(a), have almost no effect on SPDT "on" throw but improve isolation for the "off" throw, and will be introduced and studied at the end of this section.

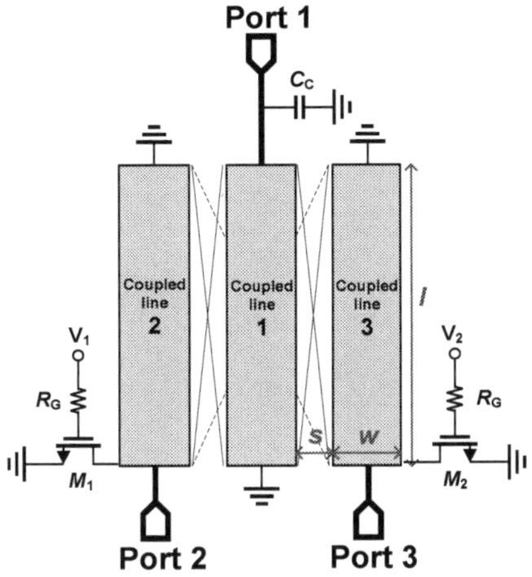

(a)

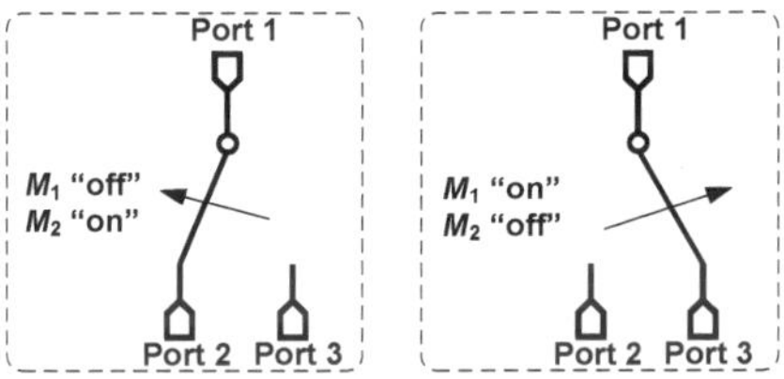

(b)

Figure 2.3 Proposed SPDT switch using artificial resonator: (a) configuration; (b) two operation modes.

Figure 2.3(b) shows the two operation modes of the switch. In the operation mode when transistor M_2 is turned on, it creates low impedance at Port 3 which isolates Port 3 from Port 1. Meanwhile, transistor M_1 is turned off, whose parasitic components and coupled-lines form an artificial resonant network with a bandpass transmission response from Port 1 to Port 2. In the following, the shunt transistors and the resonant network are analyzed in detail with small-signal equivalent circuit models.

2.2.1 Analysis of nFET Switch Transistor in Shunt Configuration

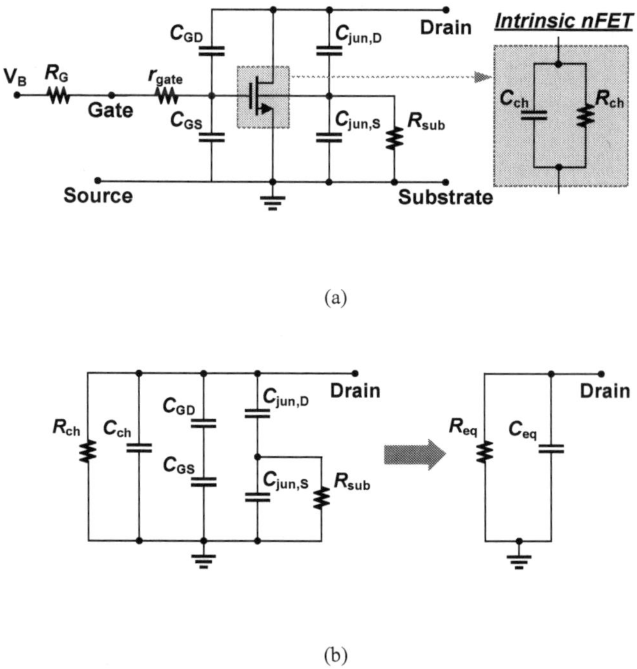

Figure 2.4 Small-signal equivalent circuit model of nFET switch transistor: (a) transistor intrinsic and extrinsic components; (b) simplified model of shunt transistor.

In Figure 2.4, the shunt n-type field-effect-transistor (nFET) switch transistor is analyzed using a small-signal circuit model, which comprises an intrinsic FET [44] and extrinsic/parasitic components [45]. Since the transistor is configured in shunt connection, its intrinsic part is modeled as C_{ch} and R_{ch} according to [36] and [44], where C_{ch} is the channel capacitance and R_{ch} is the channel resistance as shown in Figure 2.4(a). The extrinsic components are modeled as the gate-drain capacitance C_{GD}, the gate-source capacitance C_{GS}, the drain-bulk junction capacitance $C_{jun,D}$, the source-bulk junction capacitance $C_{jun,S}$, the resistance to substrate R_{sub}, and the gate resistance r_{gate}. In most of the mm-wave switch designs including this work, a large resistor R_G or a $\lambda_g/4$ T-line must be added at the gate terminal to effectively float the transistor gate terminal and reduce RF signal leakage. Thus, the model is reduced into Figure 2.4(b), where the equivalent shunt resistance and capacitance are derived in (2.1) and (2.2) as follows:

$$R_{eq} = \frac{R_{ch}\left(4R_{sub}^2\omega^2C_j^2+1\right)}{\left(4+\frac{R_{ch}}{R_{sub}}\right)R_{sub}^2\omega^2C_j^2+1}, \tag{2.1}$$

$$C_{eq} = C_{ch} + \frac{C_j}{2} + \frac{C_j\left(2R_{sub}^2\omega^2C_j^2+1\right)}{4R_{sub}^2\omega^2C_j^2+1}, \tag{2.2}$$

where $C_j = C_{jun,S} = C_{jun,D}$.

The equivalent shunt resistance in (2.1) has two expressions depending on the gate bias voltage V_B. In the condition $V_B = V_{DD}$, the transistor operates in triode region and $R_{ch\_ON} \ll R_{sub}$. Thus, this equivalent resistance is approximated in (2.3) as follows:

$$R_{\mathrm{eq\_ON}} \approx \frac{R_{\mathrm{ch\_ON}}\left(4R_{\mathrm{sub}}^{2}\omega^{2}C_{\mathrm{j}}^{2}+1\right)}{\left(4+0\right)R_{\mathrm{sub}}^{2}\omega^{2}C_{\mathrm{j}}^{2}+1} = R_{\mathrm{ch\_ON}}, \tag{2.3}$$

which is frequency-independent, but relates to the transistor size and process characteristics. In another condition $V_B = 0$ V, transistor is off with $R_{ch\_OFF} >> R_{sub}$, and the shunt resistance becomes frequency-dependent as

$$R_{\mathrm{eq\_OFF}} \approx \frac{R_{\mathrm{ch\_OFF}}\left(4R_{\mathrm{sub}}^{2}\omega^{2}C_{\mathrm{j}}^{2}+1\right)}{\frac{R_{\mathrm{ch\_OFF}}}{R_{\mathrm{sub}}}R_{\mathrm{sub}}^{2}\omega^{2}C_{\mathrm{j}}^{2}+1} \approx \begin{array}{l} R_{\mathrm{ch\_OFF}}\big|_{\text{at low frequency}} \\ 4R_{\mathrm{sub}}\big|_{\text{at high frequency}} \end{array}. \tag{2.4}$$

In Figure 2.5(a), the equivalent shunt resistances in both conditions are plotted based on transistors from GLOBALFOUNDRIES (GF) 65-nm bulk CMOS technology. As predicted, the R_{eq_ON} remains about 6 Ω from 20 to 200 GHz, while the R_{eq_OFF} is >1 kΩ at 20 GHz but decreases rapidly to only 160 Ω at 200 GHz. Generally in switch designs, the smaller R_{eq_ON} of shunt transistor leads to better isolation as it creates lower impedance at drain terminal. The R_{eq_OFF} directly influences the insertion loss of switches, as the finite impedance at drain terminal presents a leakage path for RF signals. Thus, switches using the same technology tend to have higher insertion loss when operated at higher frequencies.

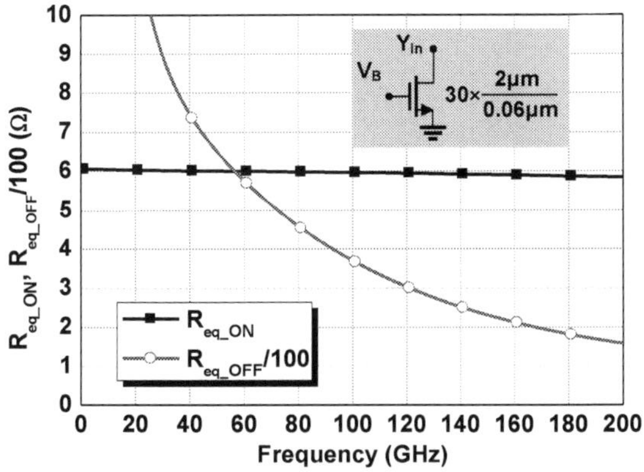

(a)

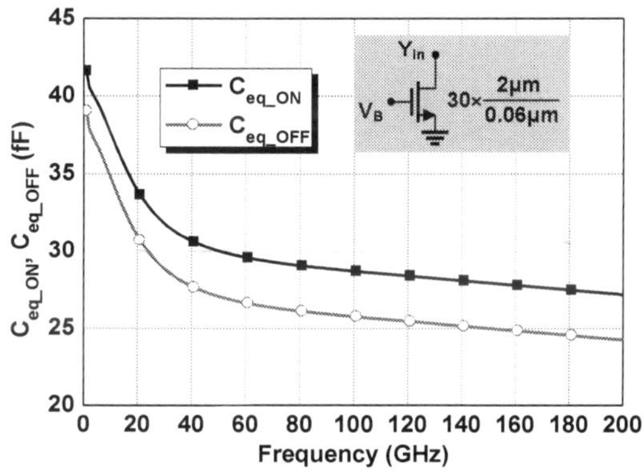

(b)

Figure 2.5 Simulated results of (a) equivalent shunt resistances for $V_B = 0/V_{DD}$, (b) equivalent shunt capacitance for $V_B = 0/V_{DD}$.

If we use R_{eq_OFF}/R_{eq_ON} as a figure-of-merit (FoM) to evaluate the shunt transistors, this process has a value of 58 at 100 GHz. In the 90-nm SiGe HBT technology [36], this FoM is up to 800 at 94 GHz that resulted in very low insertion loss with high isolation of the switch designs.

The equivalent shunt capacitance in (2.2) is approximated to (2.5) as

$$C_{\text{eq}} \approx \frac{C_{\text{j}}}{2} + \frac{C_{\text{j}}\left(2R_{\text{sub}}^2\omega^2 C_{\text{j}}^2 + 1\right)}{4R_{\text{sub}}^2\omega^2 C_{\text{j}}^2 + 1} \approx \begin{array}{l} 1.5C_{\text{j}}\big|_{\text{at low frequency}} \\ C_{\text{j}}\big|_{\text{at high frequency}} \end{array}, \tag{2.5}$$

because $C_{\text{ch}} << C_{\text{j}}$. According to (2.5), C_{eq} decreases with frequency. It is noted that the junction capacitance C_{j} has additional contribution from gate-channel capacitances in the triode region. Thus, the $C_{\text{eq\_ON}}$ is slightly larger than $C_{\text{eq\_OFF}}$ as shown in Figure 2.5(b).

In conventional $\lambda_g/4$ T-line SPDT designs, the shunt capacitances are matched out using inductive short-stubs T_{M} as in Figure 2.2. However, these T-lines introduce additional losses. In this work, the capacitances form part of the resonant network, which will be investigated using equivalent circuit model as follows.

2.2.2 Small-signal Equivalent Circuit Model of Artificial Resonator

For simplicity, the equivalent resistances of shunt switch transistors are neglected and assumed as short-/open-circuit as $R_{\text{eq\_ON}} << 50\ \Omega << R_{\text{eq\_OFF}}$ (see Figure 2.4(a)). In actuality as previously described, these resistances have determinative effects on insertion loss and isolation performance of designed switches. However, the R_{eq_OFF}/R_{eq_ON} ratio is basically a process-driven factor and is difficult to be improved with proper circuit design techniques. Thus, the following analysis focuses on the resonance property of the proposed SPDT switches, which models the switch transistors as C_{eq_ON} and C_{eq_OFF}.

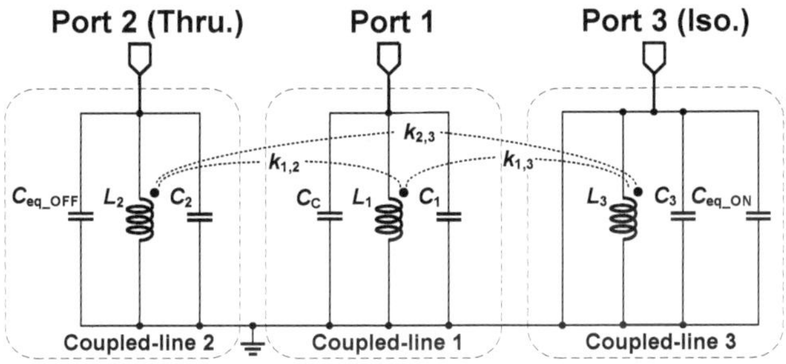

Figure 2.6 Small-signal equivalent circuit model of the proposed SPDT switch in one operation mode.

In Figure 2.6, the small-signal equivalent circuit model is built for the operation mode where Port 2 is through port and Port 3 is isolation port, i.e. $V_1 = 0$ V and $V_2 = V_{DD}$ as in Figure 2.3. The opposite mode has similar circuit model. In the model, the three coupled-lines are modelled as lump inductors L_1, L_2, L_3, and lump capacitors C_1, C_2, C_3, according to the transcendental T-line model equations as follows [46]:

$$L = L_1 = L_2 = L_3 = \frac{Z_C \tan(\beta l)}{\omega}, \tag{2.6}$$

$$C = C_1 = C_2 = C_3 \approx \frac{\varepsilon_0 \varepsilon_r w\, l}{h}, \tag{2.7}$$

where Z_C and β are the characteristic impedance and propagation phase constant of the standalone T-line with length of l and width of w, ε_r is the effective dielectric constant of the stacked dielectric materials, and h is the distance between signal line and metal ground. The magnetic coupling between coupled-lines can be extracted using electromagnetic (EM) simulator using (2.8) as

$$k_{i,j} = \frac{1}{L_i L_j} \left[\frac{\mathrm{Im}(Z_{i,j})_{\mathrm{DUT}}}{\omega} \right]_{\text{Low frequency}}. \tag{2.8}$$

As shown in Figure 2.7, the reduced small-signal equivalent circuit resembles a magnetic-coupled LC resonator, with the equivalent coupling coefficient derived as in [47] as follows:

$$k_{\mathrm{eq}} = k_{1,2} - k_{2,3} k_{1,3} = k_{1,2}\left(1 - k_{2,3}\right). \tag{2.9}$$

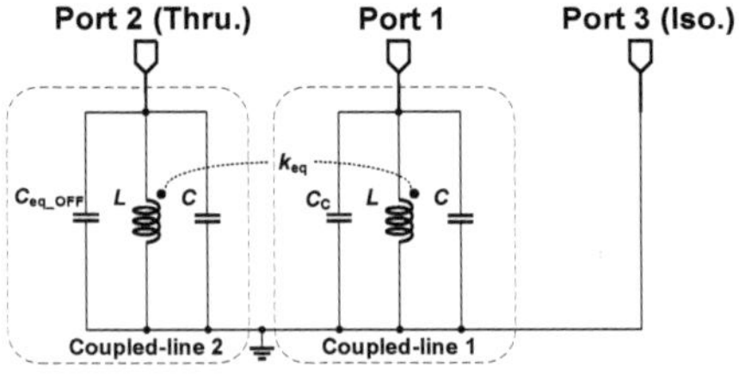

Figure 2.7 Reduced small-signal equivalent circuit model from Figure 2.6.

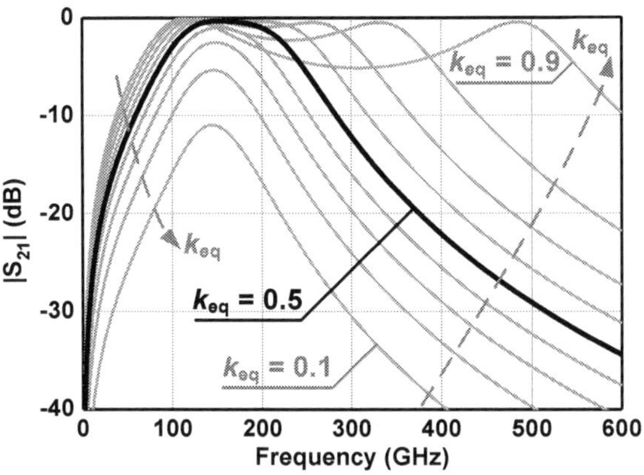

Figure 2.8 Simulated transmission responses for k_{eq} = 0.1 to 0.9 in step of 0.1.

Thus, a second-order bandpass frequency response from Port 1 to Port 2 is expected. The transmission responses are simulated by assuming L = 35 pH, C = 5 fF, C_{eq_OFF} = 20 fF, and C_C = 30 fF, and plotted in Figure 2.8 for different equivalent coupling coefficients. It is noticed that the two transmission poles are separated apart with stronger coupling, while become closer by reducing the coupling strength. The optimum response is achieved at around k_{eq} = 0.5 for this example.

Further, the locations of the two transmission poles which are the resonance frequencies can be derived in (2.10) as follows:

$$f_{P1,P2} = \sqrt{\frac{2}{f_1^2 + f_2^2 \pm \sqrt{(f_1^2 - f_2^2)^2 + 4k_{eq}^2 f_1^2 f_2^2}}} \cdot f_1 f_2, \tag{2.10}$$

where the f_{P1} and f_{P2} denote the first and second resonance poles with

$$f_1 = \frac{1}{2\pi\sqrt{L(C_{\text{eq\_OFF}} + C)}}, \tag{2.11}$$

$$f_2 = \frac{1}{2\pi\sqrt{L(C_{\text{C}} + C)}}. \tag{2.12}$$

The two transmission poles are calculated and plotted in Figure 2.9, which are well agreed with the transmission responses in Figure 2.8.

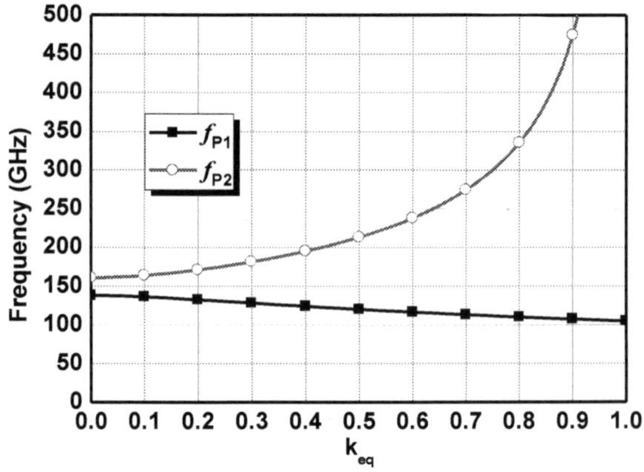

Figure 2.9 Calculated transmission poles for the design in Figure 2.8.

2.2.3 Auxiliary Coupled-lines for Isolation Enhancement

In the above investigation, the coupling coefficients between adjacent coupled-lines are assumed to be equal, i.e. $k_{1,2} = k_{1,3}$ which represents the symmetry of the three coupled-lines.

However, it is surmised that the switch isolation can be improved by reducing the magnetic coupling to the couple-line connected to isolation port, i.e. $k_{1,3}$ in the above example. Assuming L = 35 pH, C = 5 fF, C_{eq_OFF} = 20 fF, C_C = 30 fF, $k_{1,2}$ = 0.5, $k_{2,3}$ = 0.15, switch performance is studied with $k_{1,3}$ reduced from 0.5 to 0.4 in step of 0.02. In addition, R_{eq_ON} = 5 Ω is added in parallel to Port 3 in order to get meaningful isolation results. In Figure 2.10, it is observed that by reducing $k_{1,3}$ from 0.5 to 0.4, the insertion loss almost remains the same while the isolation is improved by 3 to 4 dB.

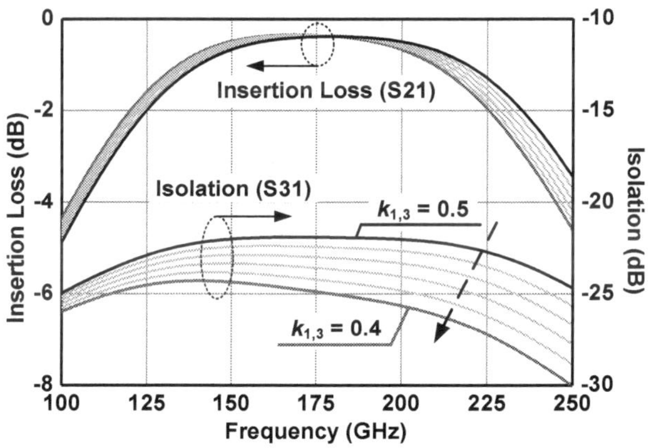

Figure 2.10 Isolation investigation by reducing coupling coefficient $k_{1,3}$.

Therefore, in the proposed SPDT switch, switchable auxiliary coupled-line is inserted between main coupled-lines to control the magnetic coupling between them, as shown in Figure 2.11(a). Switch transistors are placed in series at each end of the auxiliary coupled-line. The two operating states are depicted in Figure 2.11(b). In the state $V_a = 0$ V, the equivalent capacitance of the switches present impedances much larger than series resistance of the auxiliary coupled-line, which makes the auxiliary coupled-line a floating line. Thus, the effective coupling coefficient between main coupled-lines $k_{1,2\text{-eff}}$, is approximately equal to $k_{1,2}$. In the state $V_a = V_{DD}$, the small resistance of switches short the auxiliary coupled-line to ground. Subsequently, the effective coupling coefficient $k'_{1,2\text{-eff}}$ is approximated to $k_{1,2} - k_{1,a}k_{2,a}$.

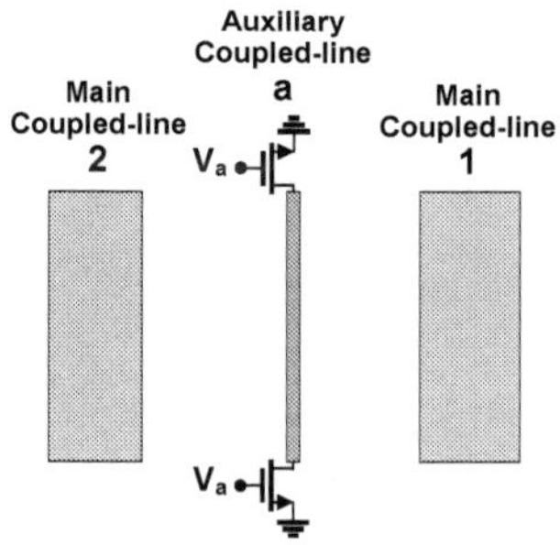

(a)

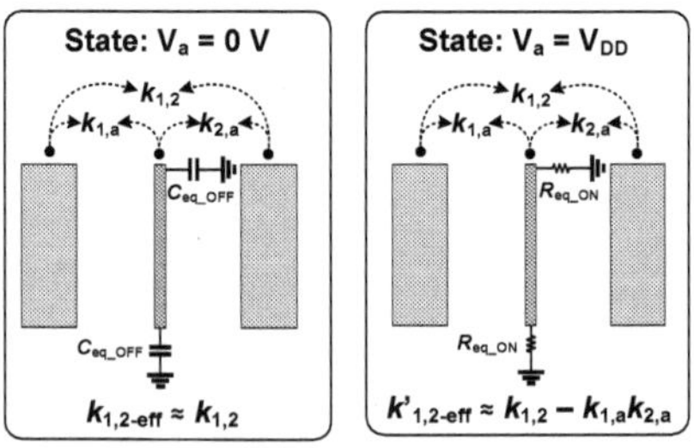

(b)

Figure 2.11 Switchable auxiliary coupled-line between main coupled-lines: (a) schematic; (b) two operating states.

To this end, the complete configuration of the proposed SPDT switch by using magnetically switchable artificial resonator is shown in Figure 2.12. In operation mode with Port 2 as signal through port and Port 3 as isolation port, the control voltages are set as follows: $V_1 = V_a = 0$ V and $V_2 = V_b = V_{DD}$. Another operation mode can be set with alternated control voltages.

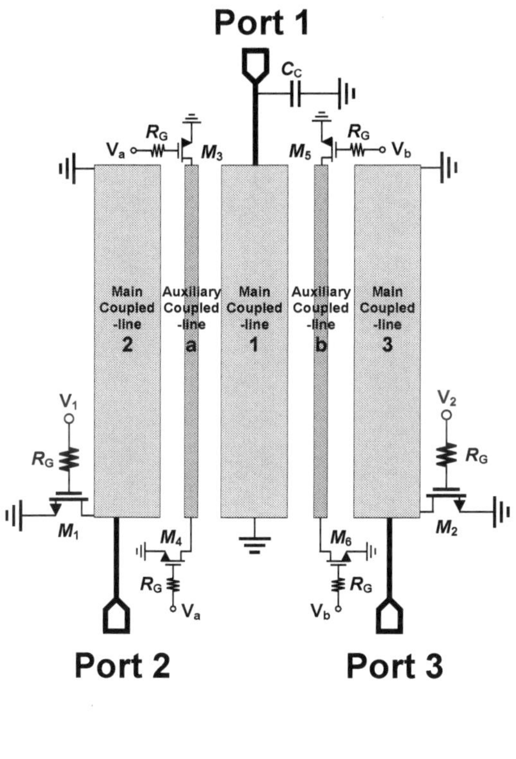

Figure 2.12 Configuration of the proposed SPDT switch using magnetically switchable artificial resonator.

2.3 130-180 GHz SPDT Switches

2.3.1 Design and Implementation

The SPDT switch is designed based on GLOBALFOUNDRIES (GF) 65-nm bulk CMOS technology. Eight copper layers are available, where four layers are utilized in the design as shown in Figure 2.13. The bottom layers M1 and M2 are stacked to form ground plane with low resistivity for the coupled-lines. The top copper layer M8 has thickness of 3.3 μm and is used as the signal lines of main coupled-lines. The second top layer M7 with thickness of 0.9 μm is used as the signal lines of auxiliary coupled-lines. The EM simulations are performed using ANSYS HFSS v.15.

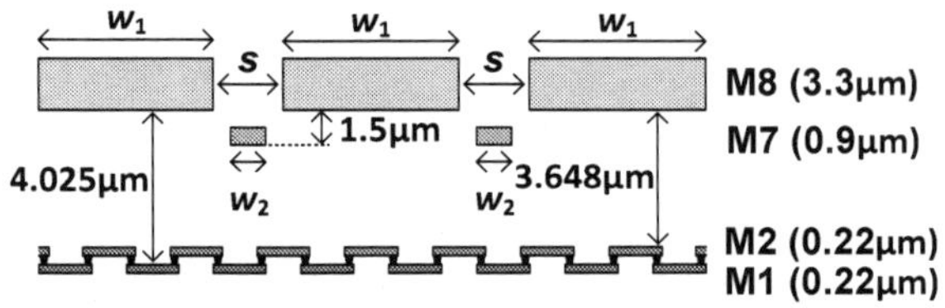

Figure 2.13 Circuit diagram of the stacked cell.

In this work, the SPDT prototype is targeted to operate at 160 GHz and achieve minimum insertion loss with isolation better than 20 dB. As discussed previously, larger transistor sizes with lower R_{eq_OFF} and R_{eq_ON} will benefit switch isolation performance but degrade the insertion loss. Thus, it becomes necessary to determine the correct transistor size at first design step. However, unlike switch designs at lower operating frequencies, it is difficult to theoretically estimate the switch performance with good accuracy beyond 100 GHz. Therefore, in this design, we test several transistor sizes and choose the optimum design with isolation just above 20 dB for the fabrication. In the following, the design procedures for the fabricated switch are illustrated in details:

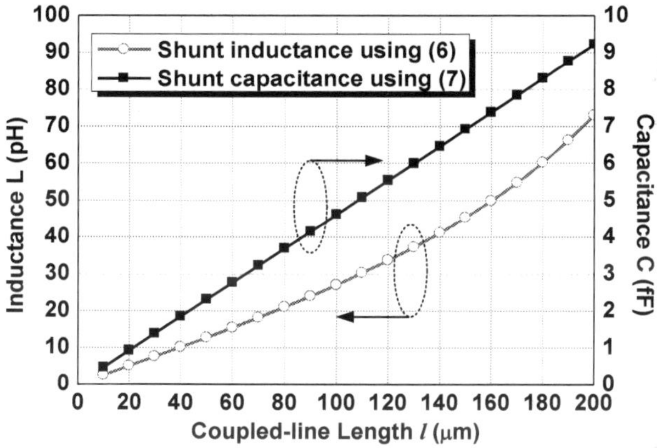

Figure 2.14 Calculated shunt inductances and capacitances of the coupled-lines versus coupled-line lengths.

1. Assume transistor size of 60 μm/0.06 μm. From Figure 2.5, C_{eq_OFF} is estimated to be 25 fF.

2. Width of main coupled-lines is determined as w_1 = 5 μm using EM simulations to obtain characteristic impedance of 50 Ω as in Figure 2.13. It is convenient to use the same width T-line to connect coupled-lines and testing pads.

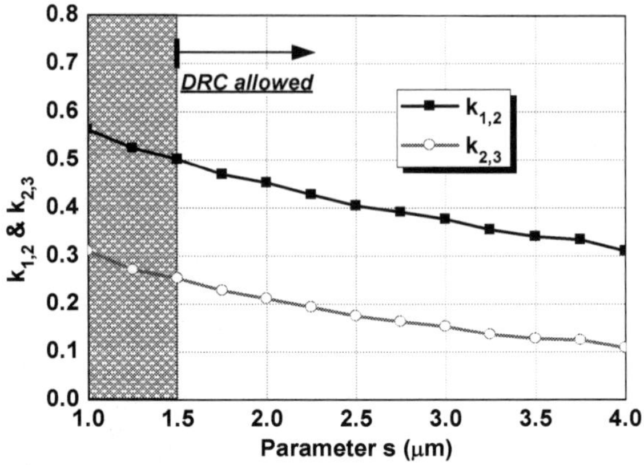

Figure 2.15 Extracted coupling coefficients between main coupled-lines versus coupled-line space *s*.

3. Shunt inductance L and capacitance C of coupled-lines are calculated using (2.6) and (2.7), which are plotted in Figure 2.14. The self-resonance frequency f_1 in (2.11) is approximated to be 160 GHz. From (2.11) and Figure 2.14, the length of coupled-lines l is estimated to be 115 μm.

4. Coupling coefficients are extracted by varying coupled-lines space s. As shown in Figure 2.15, space s = 1.5 is chosen to provide proper transmission response as indicated in Figure 2.8.

5. The whole structure is EM simulated and optimized with foundry transistor models, as shown in Figure 2.16. Width of auxiliary coupled-lines is determined as w_1 = 1 μm to obtain good shielding effects without affecting coupling between main coupled-

lines. The optimized device sizes and dimension parameters are summarized in Table I.

Table I Optimized Device Sizes and Dimension Parameters.

Transistor W/L ($M_{1\text{-}6}$)		C_C	R_G
30×(2 μm/0.06 μm)		34.7 fF	8 kΩ
l	w_1	w_2	s
100 μm	5 μm	1 μm	2 μm

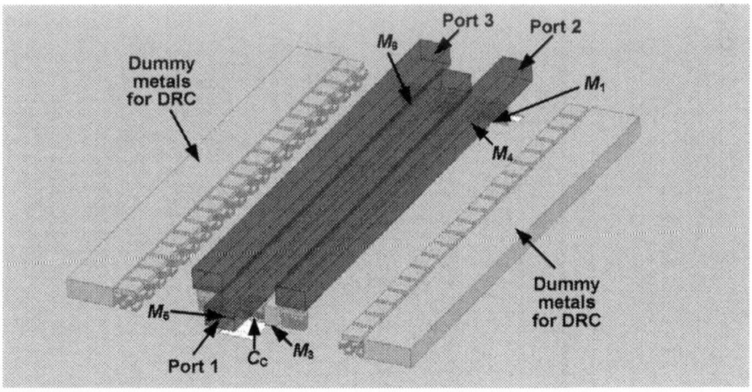

Figure 2.16 EM simulation structure in HFSS (testing pads are removed).

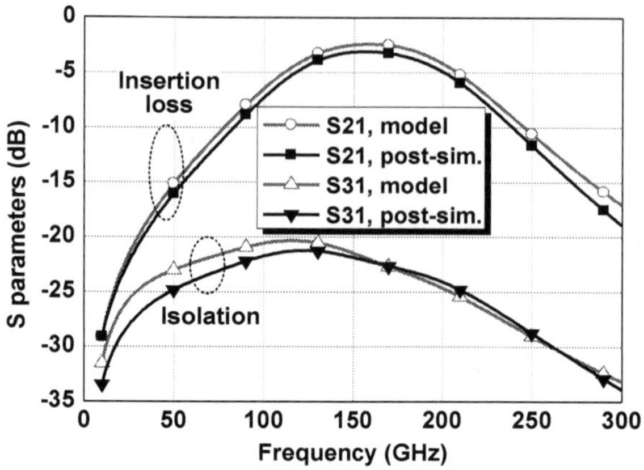

Figure 2.17 Insertion loss and isolation performance of post-simulation and circuit model. (model parameters: L = 32 pH, C = 4.6 fF, R_{eq_ON} = 5.9 Ω, R_{eq_OFF} = 230 Ω, C_{eq_ON} = 28 fF, C_{eq_OFF} = 25 fF, C_C = 34.7 fF, $k_{1,2}$ = 0.441, $k_{1,3}$ = 0.352, $k_{2,3}$ = 0.221; R = 660 Ω is added in parallel to $L_{1\text{-}3}$ to have typical Q-factor of 20 for coupled-lines)

In the prototype, the optimized length of coupled-lines l = 100 μm that is only about $\lambda_g/10$ at 155 GHz, which leads to the compact size and low insertion loss of the designed switches. From the post-simulations, it is found that $k_{1,2}$ = 0.442, $k_{1,2\text{-eff}}$ = 0.441, and $k'_{1,2\text{-eff}}$ = 0.352 with auxiliary coupled-lines, which agrees with our theoretical analysis. In Figure 2.17, the post-simulation results are in a good agreement with circuit model for operation mode that $V_1 = V_a$ = 0 V and $V_2 = V_b = V_{DD}$.

Figure 2.18 shows the micrograph of the prototypes fabricated in a 65nm CMOS process. The Switch A is the proposed magnetically switchable artificial resonator switch as configured in Figure 2.16. The Switch B, by removing the auxiliary coupled lines and their connected transistors $M_{3\text{-}6}$, is also fabricated for comparison and verification on the advantages of the magnetic switching capability. The active area of the chip, including the low metal dummy-fill area, occupies only 0.0035mm^2.

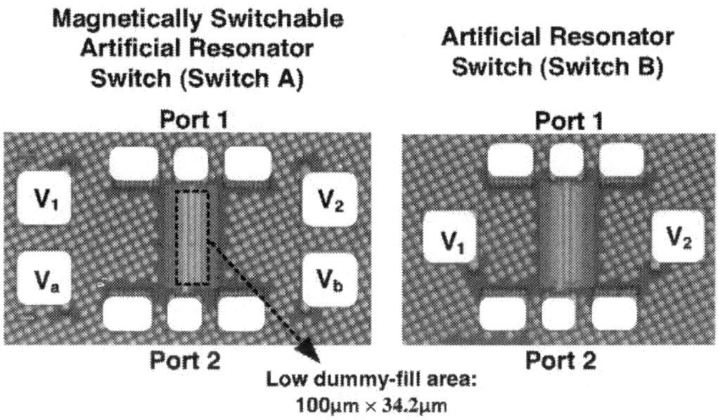

Figure 2.18 Micrograph of fabricated switches.

2.3.2 Measurements

Two-port S-parameter measurements are performed from 140 to 220 GHz using an Agilent 67-GHz PNA-X, a 140-to-220 GHz VDI's extension module and Cascade WR-5 probes, while Port 3 is internally terminated with an on-chip 50Ω resistor. The system is calibrated using a standard SOLT probe-tip calibration on a Cascade ISS substrate. Therefore, the RF pad loss is included in the insertion loss measurements.

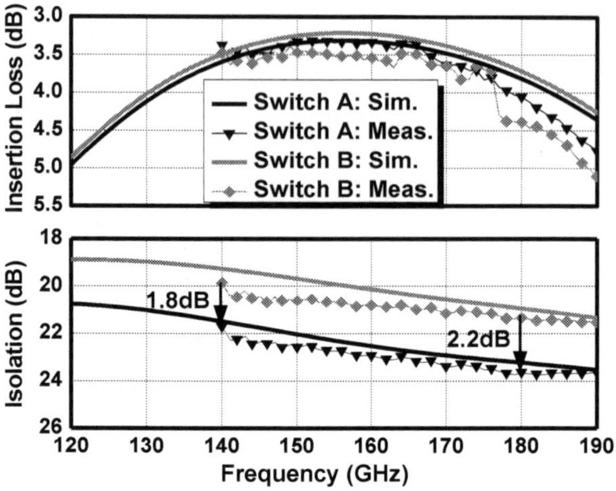

Figure 2.19 Insertion loss and isolation performance of the SPDT switches.

In Figure 2.19, the measured insertion loss and isolation of the switches are plotted which are well agreed with those of the simulations. The measured insertion loss of Switch A has little difference from the one of Switch B, which indicates the insertion loss contributed by the auxiliary coupled lines and associated transistors $M_{3\text{-}6}$ is negligible. The measured Switch A has a minimum insertion loss of 3.3 dB at 155GHz. The insertion loss curve is flat and better than 4dB from 140 to 180 GHz. By employing auxiliary coupled-lines, the isolation of Switch A is enhanced by ~2 dB as compared to the one of Switch B. Switch A has measured isolation better than 21.8 dB from 140 to 180 GHz. Using simple curve extrapolation based on the results, it is expected that the fabricated Switch A can achieve insertion loss better than 4 dB and isolation better than 21.1 dB from 130 to 140 GHz, which is not covered due to measurement setup limits. In Figure 2.20, the return losses of Switch A are better than 10 dB from 130 to 180 GHz.

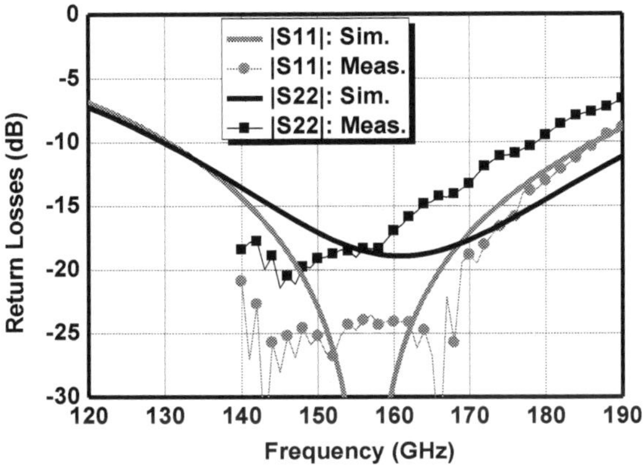

Figure 2.20 Return loss of Switch A.

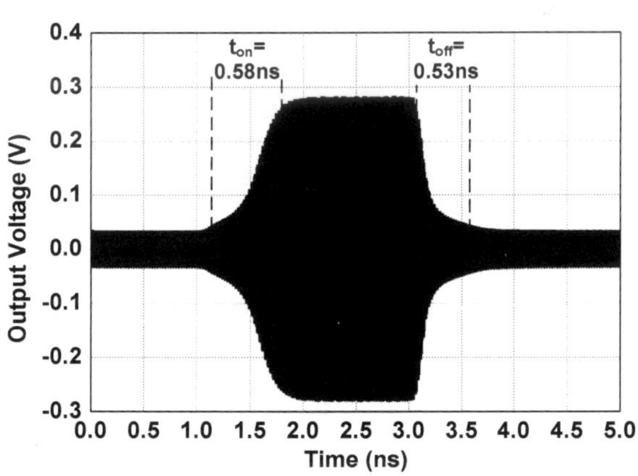

Figure 2.21 Switching-speed performance.

The switching-speed performance is evaluated using transient simulation by injecting a 0 dBm signal at 155 GHz as shown in Figure 2.21. The "on" and "off" of the 10%-to-90% switching times are 0.58 ns and 0.53 ns, respectively. Thus, the SPDT switch can support Gb/s modulation and short-pulsed active radar applications.

In the measurement setup, the output power from the VDI's extension module is below -10 dBm, which is far to saturate the design switches. Therefore, power performance is simulated with an output P_{1dB} of 11.4 dBm at 155 GHz.

Table II Performance Summary and Comparison with State-of-the-art SPDT Switches.

	[35] JSSC 2010	[48] IMS 2015	[29] JSSC 2012	[36] TMTT 2014	[37] MWCL 2014	[38] MWCL 2012	This Work
Technology f_T (GHz)	90nm CMOS 130	90nm CMOS 130	0.18μm SiGe BiCMOS 200	90nm SiGe HBT	0.13μm SiGe HBT 300	45nm SOI 485	65nm Bulk CMOS 220
Transistor $R_{ON}\times C_{OFF}$ (fs)	180	180	132	Not Mentioned	83.7	75.6	147
Frequency (GHz)	50-67	40-110	75-110	73-110	96-163	140-220	130-180
Insertion Loss (dB)	1.5-2 #	<4	2.8-3.1 #	1.1 #	2.6-3	3-4.5	3.3-4
Return Loss (dB)	>8	>10	>10	>10	>10	>10	>10
Isolation (dB)	25-27.5	>20	21-22	22	23.5-29	20-30	21.1-23.7
P_{1dB} (dBm)	13.5 †	10	Not Mentio	17	17 @ 94	10 †	11.4 †

			ned		GHz		
Size Excluding Pads (mm^2/ $1000\times\lambda_0^2$) *	0.27/ 10.3	0.11/ 2.55	0.114/ 10.8	0.213/ 19.8	0.228/ 42.5	0.10/ 36	0.0035/ 0.94
P_{DC} (mW)	0	0	0	5.9	6	0	0
Topology *	λg/4 TL + SS-SPST	Traveling-wave	λg/4 TL + SS-SPST	λg/4 TL + DS-SPST	λg/4 TL + DS-SPST	λg/4 TL + DS-SPST	Magnetically Switchable Artificial Resonator

* TL: Transmission line; SS: Single-shunt; DS: Double-shunt
\# Pad loss de-embedded † Simulation

Table II provides the performance summary and comparison with state-of-the-art SPDT switches [29, 35-38]. Compared with the CMOS designs in [29] and [35], the reported switch achieves much higher operating frequency with similar insertion loss and isolation performance. Compared with implementations using SOI and HBT technologies in [37] and [38], this work demonstrates comparable insertion loss and isolation using much worse switch transistors. Most importantly due to the advantage of proposed topology, the switch only occupies an area of 0.0035 mm^2 which is only 1.5 to 3.5% of the area occupation in previous works [29, 35-38].

2.4 A 220-285 GHz SPDT Switch

2.4.1 Design and Implementation

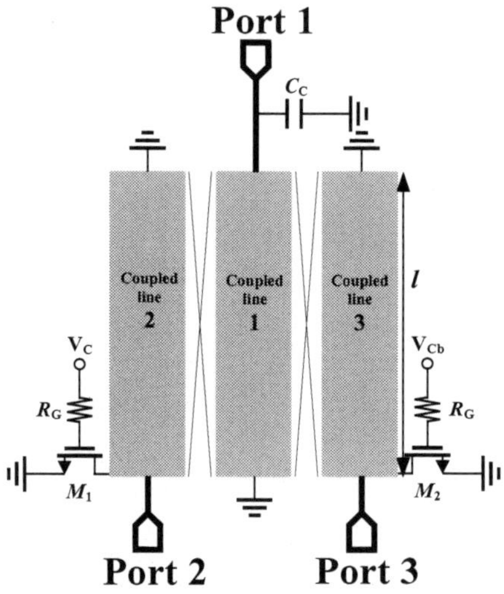

(a)

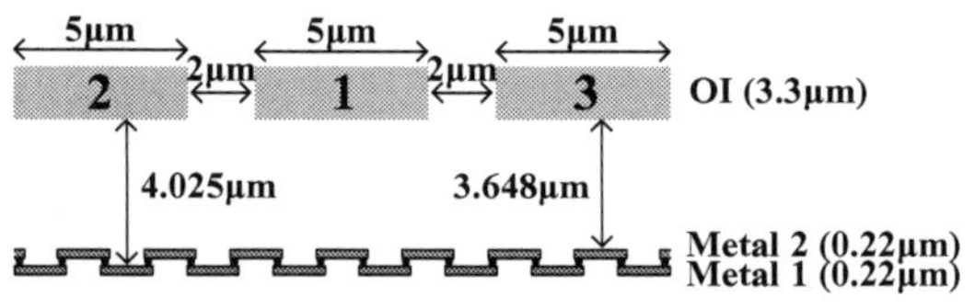

(b)

Figure 2.22 Proposed SPDT switch using the switchable resonator concept: (a) Configuration; (b) Cross-section view. (Transistors M_1 and M_2: 17 fingers, single finger W = 2μm, L = 0.06μm; C_C = 19.9 fF; R_G = 8 kΩ; l = 60 μm)

In Figure 2.22, the proposed SPDT switch is designed in a commercial 65-nm bulk CMOS. It comprises three coupled- lines with length l, two normal RF nFET transistors M_1 and M_2, one compensating capacitor C_C, and two biasing resistors R_G at transistor gate terminals.

A 220-285 GHz SPDT switch is designed using this topology. Switch transistors are implemented using normal RF nFET with $R_{ON}\times C_{OFF}$ of 147 fs, while coupled-lines are built using Metal OI as signal lines and stacked Metal 1 and Metal 2 as ground plane with low resistivity. The design parameters are optimized using 3-D full-wave EM simulations by HFSS and circuit co-simulations, which are depicted in Figure 2.22(b) and caption of Figure 2.22. The fabricated SPDT switch is shown in Figure 2.33, where the active area is only 0.002 mm^2.

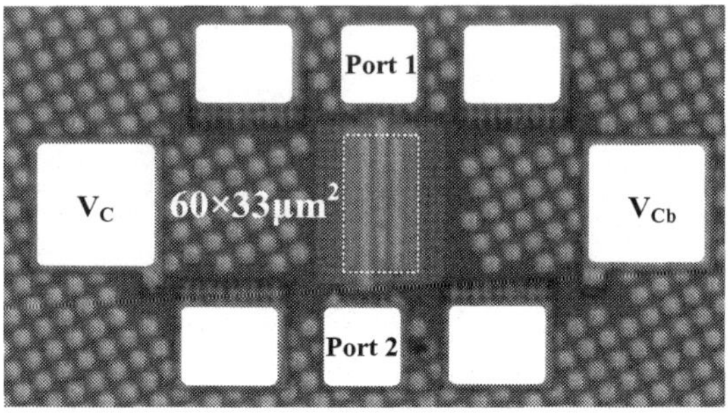

Figure 2.23 Die micrograph of fabricated switch.

2.4.2 Measurements

Two port S-parameter measurements are performed from 220 to 320 GHz using an Agilent 67-GHz PNA-X, a 220-320 GHz VDI's extension module and Cascade WR-3 probes, while Port 3 is internally terminated with an on-chip 50Ω resistor. The system is calibrated using an SOLT probe-tip calibration on a Cascade ISS substrate. Therefore, the RF pad loss that is 0.5-0.7 dB for each pad at 250 GHz is included in the insertion loss measurements.

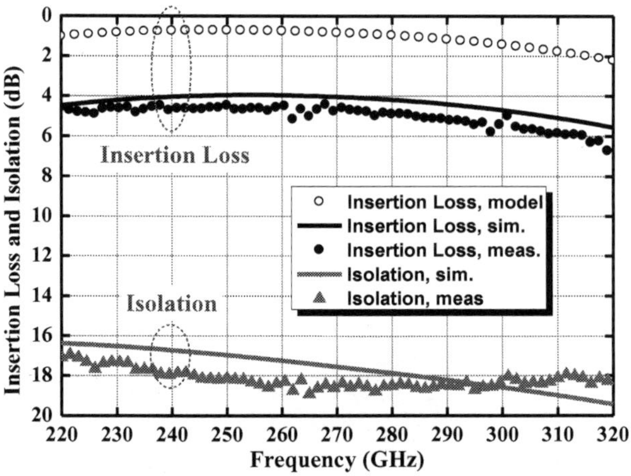

Figure 2.24 Insertion loss and isolation from 220 to 320 GHz. (Model in Figure 2(b): $L_1 = L_2 =$ 19 pH, $C_{OFF} + C_2 = 17.4$ fF, $C_C + C_1 = 24$ fF, $k_{12\text{-}eq} = 0.4$)

Figure 2.24 shows that the measured insertion loss and isolation performance are in a good agreement with simulated ones. The measured switch has a minimum insertion loss of 4.2 dB at 250 GHz, including the pad losses of 1.0-1.4 dB. The insertion loss curve is flat with value less than 5 dB from 220 to 285 GHz. The measured isolation curve is also flat with value around 18 dB. The simulated isolation between Port 2 and Port 3 is better than 18 dB. Thus, the switch has ON/OFF ratio of better than 13 dB that satisfies the terahertz imaging requirements [4].

Figure 2.25 shows the return losses that are better than 10 dB from 220 to 300 GHz. The simulated switching ON and OFF times are 0.29 ns and 0.26 ns respectively as shown in Figure 2.26, which could support Gb/s modulation and short-pulsed active radar applications. Our measurement setup has maximum output power from extension module of less than -10 dBm that is well below the P_{1dB} of tested switch. The simulated output P_{1dB} is 9.2 dBm at 260 GHz.

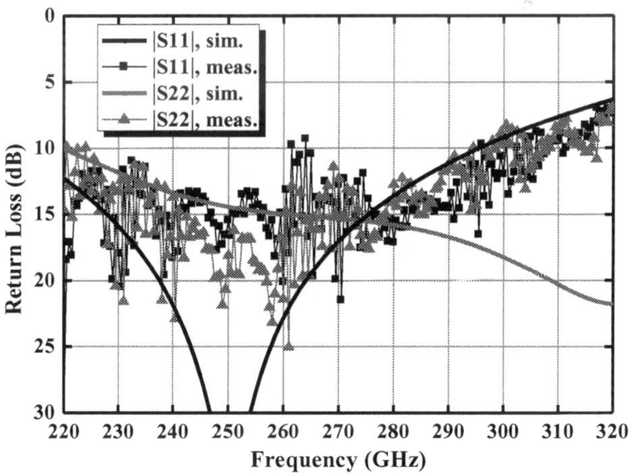

Figure 2.25 Return losses from 220 to 320 GHz.

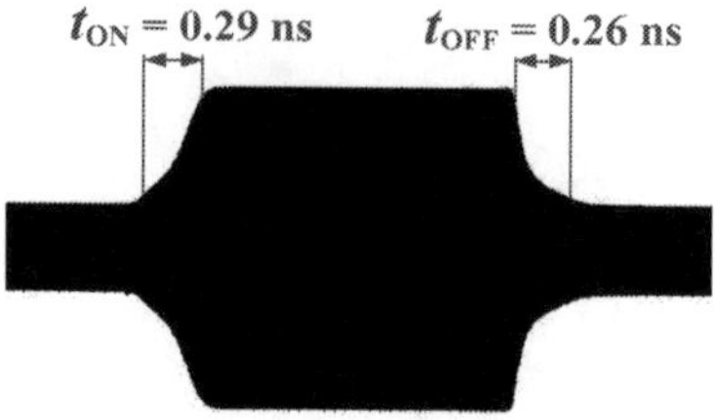

Figure 2.26 Simulated switching speed performance.

Table III shows the performance comparison of this work with the state of the art. The proposed switch features insertion loss of 4.2 dB and isolation of 19 dB that are comparable to other works. However, it achieves the highest operating frequency and smallest chip size (in terms of both mm^2 and λ_0^2 where λ_0 is the free space wavelength at center operating frequency) among SPDT switches operating beyond 100 GHz in silicon technologies.

Table III Performance Summary and Comparison of State-of-the-Art SPDT Switches Operating Beyond 100 GHz

	[49] TMTT 2010	[48] IMS 2015	[34] SiRF 2013	[37] MWCL 2014	[38] MWCL 2012	This Work
Technology f_T (GHz)	0.12-μm SiGe BiCMOS 200	90nm CMOS 130	0.12-μm SiGe BiCMOS 180	0.13μm SiGe HBT 300	45nm SOI 485	65nm Bulk CMOS 220
Transistor R_{ON}×C_{OFF} (fs)	-	180	-	83.7	75.6	147
Frequency (GHz)	85-105	40-110	110-140	96-163	140-220	220-285
Insertion	2.3-3 #	<4	2.5-3.5 #	2.6-3	3-4.5	4.2-5

Loss (dB)						
Return Loss (dB)	>10	>10	>10	>10	>10	>10
Isolation (dB)	21	>20	14.5-17.5	23.5-29	20-30	17-19
P_{1dB} (dBm)	-	10	-	17 @ 94 GHz	10 †	9.2 †
Size Excluding Pads (mm^2/ $1000\times\lambda_0^2$) *	0.0475/ 4.5	0.11/ 2.55	-	0.228/ 42.5	0.10/ 36	0.0035/ 0.94
P_{DC} (mW)	0	0	-	6	0	<4×10^{-8}
Topology *	λg/4 TL + SS-SPST	Traveling -wave	λg/4 TL + SS-SPST	λg/4 TL + DS-SPST	λg/4 TL + DS-SPST	Switchabl e Resonator

* TL: Transmission line; SS: Single-shunt; DS: Double-shunt

\# After de-embedding pad loss † Simulation - Not mentioned

2.5 Summary

This chapter demonstrated SPDT designs using the magnetically switchable artificial resonator topology. Firstly, the used CMOS transistors in shunt configuration were studied with equivalent models. Then, the switch was analyzed as three-port network formed by lump components. Its resonant property and transmission response were investigated theoretically. At last, auxiliary coupled-lines were introduced to switch the magnetic coupling strength between main coupled-lines to improve the switch isolation.

Two SPDT prototypes operating at 130 to 180 GHz and one SPDT switch operating at 220 to 285 GHz were designed and fabricated in a commercial 65-nm CMOS technology. The switch obtains state-of-the-art performance even comparing to the SOI and HBT implementations, with tremendous reduced circuit size. The compact size saves the implementation cost of communication systems especially in large-scale arrays.

CHAPTER 3

60-GHz Digitally-Controlled Phase Shifters

3.1 Introduction

Recently, intensive research efforts have been focused on millimeter-wave (mm-wave) beam-forming techniques for commercial applications [3, 5, 15, 16]. Despite its advantages of high directivity, beam-steering capability, and wide spatial coverage, beam-forming requires power-hungry and bulky multiple channels to increase beam coverage under high gain condition. Thus, it is crucial to develop low-power and small form-factor circuits for beam-forming systems to be adopted by commercial markets, where power and cost are major driven factors. As the key building block to control beam directions through controlling phase delays in each of the multiple channels in beam-forming systems, phase shifter is required especially to be designed with high phase-resolution, small phase/gain errors, low insertion loss, low power consumption, and compact circuit size for adoption in commercial application systems [5, 12, 13, 17, 19-21, 23, 24, 50-142].

The active vector modulation (VM) approach by adding orthogonal modulated signals using variable gain amplifiers (VGA) to achieve variable output phases was demonstrated in [20] and [66] at mm-wave frequencies. Despite its benefits of compact size and low insertion loss, the VM phase shifter still suffers from large phase error and power consumption especially at mm-wave frequency [66].

On the other hand, the passive approach for mm-wave phase shifter designs is more popular. It is composed of mainly two topologies, reflective-type phase shifter (RTPS) [12, 86, 87, 114] and switched-type phase shifter (STPS) [13, 51, 64, 71, 90, 92, 107]. In [86] and [87], compact 60-GHz RTPS designs with good phase responses and low insertion loss were presented. But total phase shift ranges were limited to 180°, because the insertion loss variation in different phase states will became too large for a 360° design. Meanwhile, STPS is built by cascading switched networks with relative phase shifts of 0/180°, 0/90°, 0/45° and etc. This topology has advantage of directly digital control [92] and is suitable for designs with large phase shifts. In [51, 64, 90], 4-bit phase shifters for the commercial 60-GHz, 77-GHz, and 94-GHz applications were demonstrated with small phase/gain errors and P1dB of larger than 10 dBm. A 60-GHz 5-bit phase shifter was proposed in [13] for enhancing the error vector magnitude performance of beam-forming systems. However, these STPSs are too bulky to be economically used in beam-forming systems.

In this chapter, a 57-64 GHz 5-bit phase shifter comprising both STPS topology and digital RTPS topology is reported. The chapter is organized as follows. Section 3.2 presents the switched-varactor reflective-type topology. Based on the investigation, a 60-GHz 3-bit 90° phase shifter is designed and verified experimentally. Section 3.3 presents the theoretical analysis of the cross-coupled bridged T-type topology. Based on the study, 60-GHz 0/180° and 0/90° are designed. Section 3.4 presents the architecture, design, and measured data of the 60-GHz 5-bit 360° phase shifter. A brief summary is drawn in Section 3.5.

3.2 Switched-Varactor RTPS

3.2.1 Theoretical Analysis

The RTPS uses a 90° hybrid coupler and two variable reflective loads [114]. Its phase shift varies according to the phase angle of reflection coefficient as

$$\Gamma_{\text{Load}} = \frac{jX - Z_0}{jX + Z_0}, \tag{3.1}$$

where X is the reactance of the reflective loads with minimum and maximum values denoted as X_{min} and X_{max}, respectively. The characteristic impedance of hybrid coupler is denoted as Z_0. Thus, the absolute output phase and total phase shift are expressed as

$$\varphi_{\text{out}} = -\pi - 2\arctan\left(\frac{X}{Z_0}\right), \tag{3.2}$$

$$\varphi_{\text{total}} = 2\left|\arctan\left(\frac{X_{\max}}{Z_0}\right) - \arctan\left(\frac{X_{\min}}{Z_0}\right)\right|. \tag{3.3}$$

Figure 3.1 shows the schematic of proposed reflective load. The reflective load is formed by N varactors (C_{V1}, C_{V2}, C_{V3}, ..., C_{VN}), one fixed-value parallel capacitor C_{Fix}, and one series inductor L_T. The biasing circuits are formed by resistors R_B and capacitors C_B that are equivalent open- and short-circuit at operating frequency.

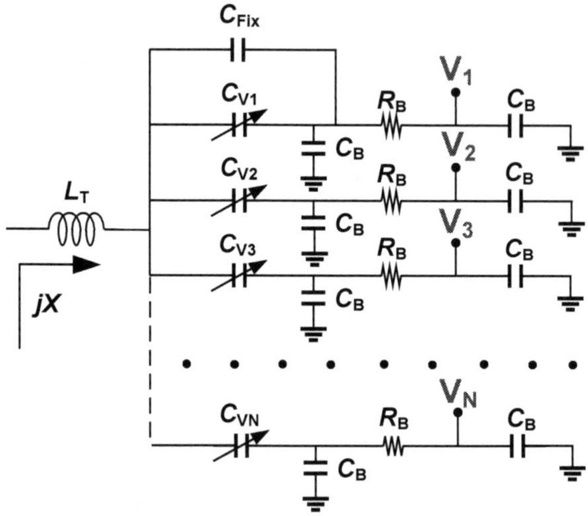

Figure 3.1 Schematic of the reflective load for an N-bit switched-varactor RTPS.

The N control bits (V_1, V_2, V_3, …, V_N) are biased at only two voltages states, i.e. 0 V and 1.2 V for off and on digital operations respectively. If we assume that the tuning ratio of varactors is equal to *r*, the varactors have only two capacitance values under alternative digital biasing voltages as

$$C_{\mathrm{Vi}} = \begin{cases} C_{\mathrm{Vi\text{-}min}} \\ r \times C_{\mathrm{Vi\text{-}min}} \end{cases}. \tag{3.4}$$

Neglecting the parasitic resistance, the reactance of proposed reflective load is derived as

$$X = \omega L_{\mathrm{T}} - \frac{1}{\omega(C_{\mathrm{Fix}} + C_{\mathrm{V1}} + C_{\mathrm{V2}} + C_{\mathrm{V3}} + \cdots + C_{\mathrm{VN}})}. \tag{3.5}$$

The varactors sizes are binary-weighted as

$$C_{\mathrm{V1\text{-}min}} = \frac{C_{\mathrm{V2\text{-}min}}}{2} = \frac{C_{\mathrm{V3\text{-}min}}}{2^2} = \cdots = \frac{C_{\mathrm{VN\text{-}min}}}{2^{\mathrm{N}-1}}, \tag{3.6}$$

so the load reactance in (2.17) is further deduced as

$$X = \omega L_{\mathrm{T}} - \frac{1}{\omega\left(C_{\mathrm{Fix}} + \sum_{i=1}^{\mathrm{N}}\left(r_i \times 2^{i-1} \times C_{\mathrm{V1\text{-}min}}\right)\right)}, \tag{3.7}$$

where r_i is defined as the biasing factor of the varactor C_{Vi}, which is either 1 or r from (3.4). Thus, the phase shifter has 2^N output phase states that are selected by N control bits.

To minimize the insertion loss and loss variation in RTPS designs, the following expression is satisfied at frequency ω_0 [117]:

$$X_{max} + X_{min} = 0\Big|_{\omega=\omega_0} . \tag{3.8}$$

So, by using (2.15), (2.19) and (2.20), the elements are solved and expressed as

$$C_{V1\text{-}min} = \frac{2Z_0 \tan\left(\frac{\varphi_{total}}{4}\right)}{\left(2^N - 1\right)\omega_0 \left(r-1\right)\left(\omega_0^2 L_T^2 - Z_0^2 \tan^2\left(\frac{\varphi_{total}}{4}\right)\right)}, \tag{3.9}$$

$$C_{Fix} = \frac{\omega_0 L_T (r-1) - Z_0 \tan\left(\frac{\varphi_{total}}{4}\right)(r+1)}{\omega_0 \left(r-1\right)\left(\omega_0^2 L_T^2 - Z_0^2 \tan^2\left(\frac{\varphi_{total}}{4}\right)\right)} . \tag{3.10}$$

It is noted that the output phases are not strictly varying in linear according to control bits based on (2.14) and (2.19), however, N-bit phase shifter can still be properly designed with acceptable phase errors. Table II summarized the design parameters and ideal phase responses at 60 GHz for 3-bit 90° phase shifters for several design examples calculated from (2.21) and (2.22), under the assumption that $r = 3$ and $Z_0 =$ 50 Ω. The RMS phase error [17] is calculated using

$$\text{RMS phase error} = \sqrt{\frac{1}{\text{N}-1} \times \sum_{i=2}^{\text{N}} \left|\varphi_{\text{error},i}\right|^2}, \tag{3.11}$$

where N = 8 is the number of total states, i denotes the state number, φ_{error} is absolute phase error of State 2-8 using the first state as the reference. It is noted that the RMS phase error reduces if larger L_{T} and C_{Fix} are used. However, it introduces more resistance from series inductor and requires smaller varactor sizes that may reduce modelling accuracy due to the fabrication process variation in CMOS.

3.2.2 60-GHz 3-bit 90° Phase Shifter

Figure 3.2 shows the schematic of the proposed 3-bit 90° phase shifter, comprising of a transformer-type directional coupler and two switched-varactor based reflective loads.

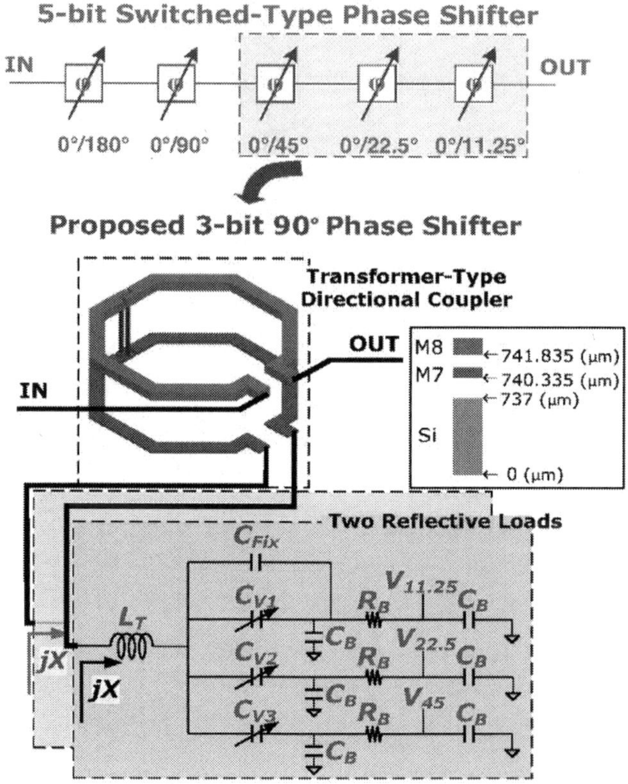

Figure 3.2 Schematic of the proposed 3-bit 90 phase shifter.

The transformer-type directional coupler is adopted for its compactness. The size reduction is over 50% comparing to the broadside coupler in [86]. As shown in Figure 3.9, the coupler uses two metal layers, colored in grey (M8) with 3.3 μm thickness and red (M7) with 0.9 μm thickness respectively. The blue traces are the metal via with length extended for better illustration. The coupler has its primary and secondary coils inter-crossed where the primary coil is firstly constructed on M8 layer for half-turn, and then is routed down to M7 layer for another half-turn to complete the winding trace. Meanwhile, the secondary coil is formed on M7 layer first, and then crosses to complete the full-turn on M8 layer. Thus, the structure is symmetrical so that the impedance looking into the four ports are expected to be the same. The optimized outer diameter is found as 69 μm with trace width of 5 μm to minimize the coupler loss while maintaining the return loss and isolation better than 15 dB. In Figure 3.9, the full-wave simulation using *ANSYS HFSS V.14* shows that the transformer-based coupler achieves the total insertion loss less than 0.9 dB and the return loss and isolation better than 17 dB over 50-70 GHz. The phase balance is kept close to 90°.

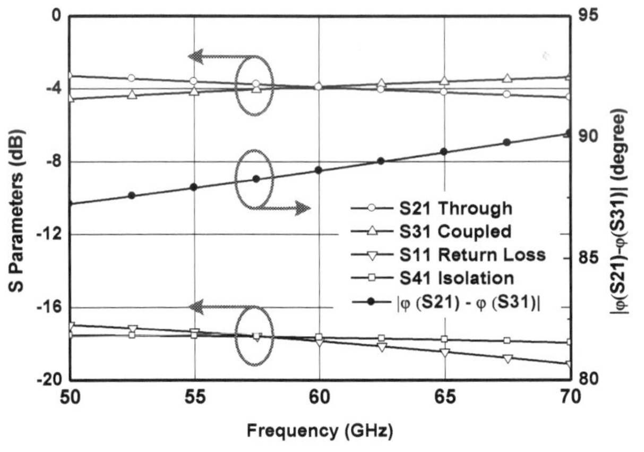

Figure 3.3 Simulated performance of the transformer-type directional coupler.

As shown in the bottom of Figure 3.3, the proposed reflective load comprises three varactors C_{V1} to C_{V3} and one fixed-value capacitor C_{Fix} in shunt-connection. The L_T is a series inductor that forms a resonator together with the collective capacitances to increase the load reactance X varying range and the phase-shifting range according to (3.3). The C_B and R_B are biasing components with values of 1 pF and 10 kΩ respectively. The digital control function is realized by biasing the three control bits, $V_{11.25}$, $V_{22.5}$ and V_{45} to either 0 V or 1.2 V. Since each varactor has two capacitance values under alternative digital biasing, the proposed reflective load is capable of achieving eight different reactance values as well as eight phase shifts for the phase shifter according to (3.2). The C_{Fix} is added to compensate the overall phase error with the tradeoff of a reduced phase-shifting range. The varactor sizes are binary-weighted and the optimized design parameters are summarized in Table IV.

In Figure 3.4, the die micrograph of the fabricated phase shifter is shown, where the circuit size excluding testing pads is only 0.034 mm^2.

Table IV Design parameters

	Fixed Components	Variable Components		
		Size (Fingers×W/L)	Min. (fF)	Max. (fF)
C_{Fix}	17.8 fF	-	-	-
$C_{V1\text{-}min}$	-	1×1.6 μm/0.47	2.47	7.4
$C_{V2\text{-}min}$	-	2×1.6 μm/0.47	4.92	14.6
$C_{V3\text{-}min}$	-	4×1.6 μm/0.47	9.78	28.7
L_T	110 pH	-	-	-

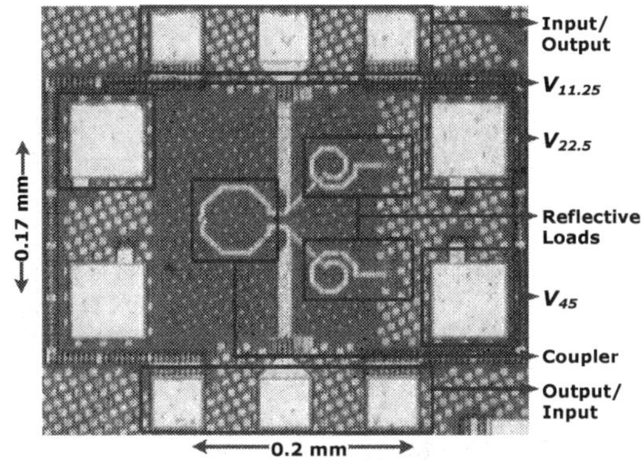

Figure 3.4 Die micrograph.

3.2.3 Measurements

The measurement is performed on-chip using Agilent N5247A PNA-X network analyzer and Cascade Elite 300 probe station.

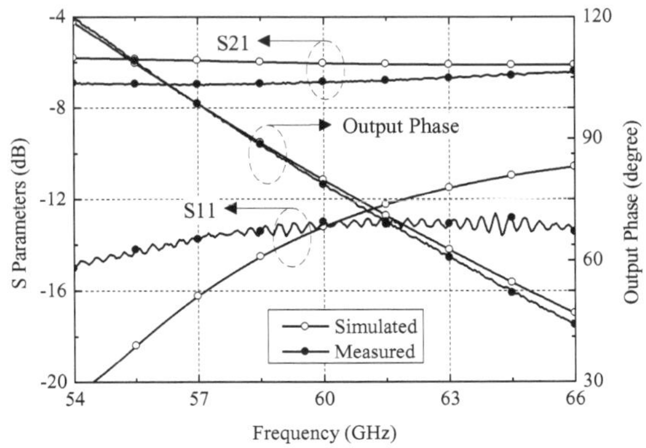

Figure 3.5 Measured and simulated performance for state 000.

Figure 3.5 compares the measured and simulated performance for state 000. Good agreement is obtained. Figure 3.6 depicts the measured phase shifts of eight states over frequency range of 54-66 GHz that fully covers the 60-GHz ISM band. The phase shifter shows the eight output phases with phase-resolution of 11.25° and an RMS phase error (calculated as in [13]) of 5.2° at 60-GHz. The measured group delays are within ±3 ps with absolute delay less than 20 ps as shown in Figure 3.6. Figure 3.7 depicts the insertion loss and return loss over the entire bandwidth. The worst loss flatness has a value of ± 0.6 dB and occurs in State 8. At 60-GHz, the average insertion loss is 5.69 dB with loss variation of ±1.22 dB across eight states.

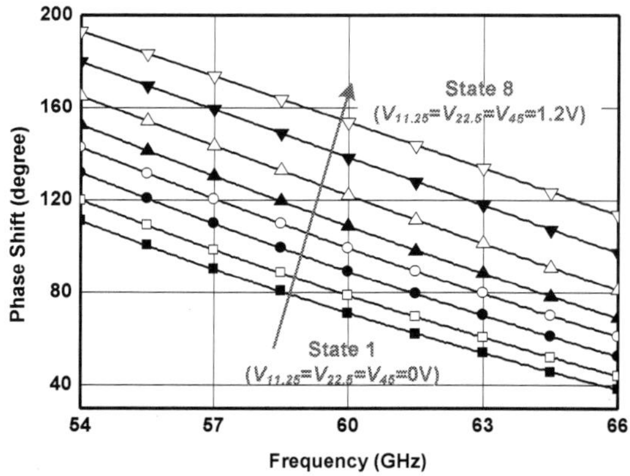

Figure 3.6 Measured phase shifts in 8 states.

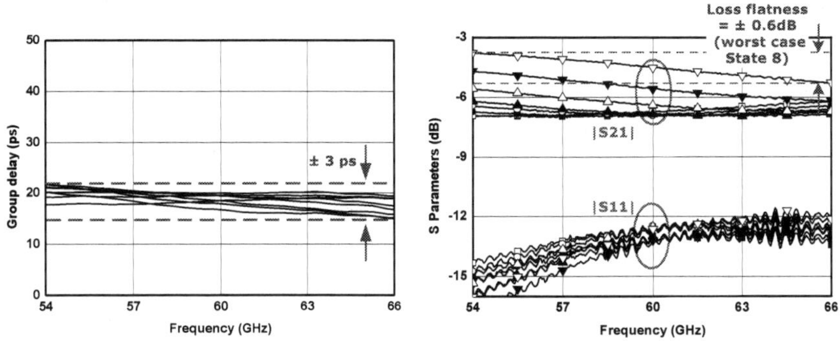

(a) (b)

Figure 3.7 (a) Measured group delays in 8 states; (b) Measured insertion loss in 8 states. (Symbols refers to Figure 3.6)

Table V compares this work with other similar phase shifters. This work has the smallest circuit size and insertion loss among 90° phase shifters. Moreover, this work is controlled directly by digital bits.

Table V Comparison with other similar phase shifters.

Reference	[5]	[86]	[13]*	This work
Load Topology/Step	RTPS/Cont's	RTPS/Cont's	STPS/3-bit	RTPS/3-bit
Technology	0.12-μm SiGe	90-nm CMOS	90-nm CMOS	65-nm CMOS
Freq. (GHz)	57-64	50-65	57-64	54-66
Phase-shifting range(°)	180	90	90	90
Insertion loss (dB)	6±1.8	6.25 ± 1.75	7.3±1.5	5.69±1.22
Return loss (dB)	-	> 12	-	> 12
DAC requirement	Yes	Yes	No	No
Size (mm^2)	0.18	0.08	0.17	0.034

* The 3-bit 90° phase shifter is used for comparison. The insertion loss is estimated by averaging the total insertion loss by the number of stages since the main loss is due to switch loss. The size is estimated from die photo.

3.3 Cross-Coupled Bridged T-type Phase Shifter

3.3.1 Theoretical Analysis

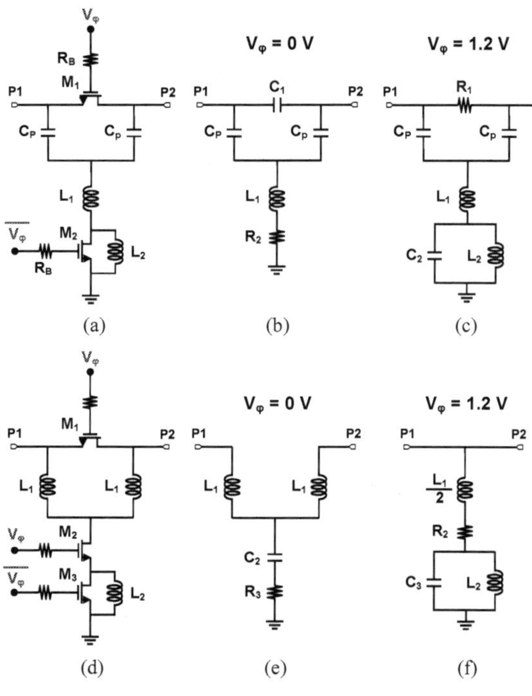

Figure 3.8 Capacitor-based bridged T-type phase shifter: (a) Topology; (b) Operation state when Vφ = 0 V; (c) Operation state when Vφ = 1.2 V. Inductor-based bridged T-type phase shifter: (d) Topology; (e) Operation state when Vφ = 0 V; (f) Operation state when Vφ = 1.2 V. (*R_1*, *R_2* and *R_3* denote the on-resistances of M1, M2 and M3; *C_1*, *C_2* and *C_3* denotes the off-capacitances of M1, M2 and M3; *R_B* is the 10 kΩ biasing resistors)

In the past, the bridged T-type topology is prevailing in mm-wave STPSs, especially in designs with small phase shifts of less than 90° [13, 51, 64, 90, 92]. In Figure 3.8(a), the capacitor-based bridged T-type phase shifter comprises two capacitors (C_p), inductors ($L_{1,2}$) and two MOS switches ($M_{1,2}$). Its phase shift is the output phase difference in the two operation states as indicated in Figure 3.8(b) and Figure 3.8(c), respectively. By properly choosing design parameters, the required phase shift can be obtained as described in [51]. However, these capacitor-based phase shifter designs generally have poor design accuracy, as the fabrication tolerance of metal-insulator-metal (MIM) and metal-oxide-metal (MOM) capacitors are typically 5-10% in advanced CMOS [15], [105]. To tackle the aforementioned issue, the two capacitors (C_p) can be replaced with inductors (L_1) that are better modeled using electromagnetic (EM) simulators as shown in the Figure 3.8(d).

Reference [107] analyzed the circuits and provided design equations under the conditions that the on-resistance and off-capacitance of transistor M_1 can be ignored as in Figure 3.8(e) and Figure 3.8(f). The equations were used in the realizations of Ku-/V-band phase shifters [13, 107]. However, this design approach has several design constraints for mm-wave phase shifters. First, the two inductors (L_1) must be separated in a large distance to prevent their mutual coupling effect modeled as coupling coefficient *k,* which is neglected in the prior arts but can dramatically degrade phase shift accuracy and performance even the *k* is small. Reducing *k* through increase the distance will lead to large silicon area as revealed in [13].

Second, off-capacitance of transistor M_1 presents significant impedance and must be considered at mm-wave frequencies. For example, the off-capacitance of M_1 in 0/90° phase shifter of [13] is ~23 fF that has an equivalent impedance of 115 Ω at 60 GHz frequency, which is not large enough to be treated as an open-circuit.

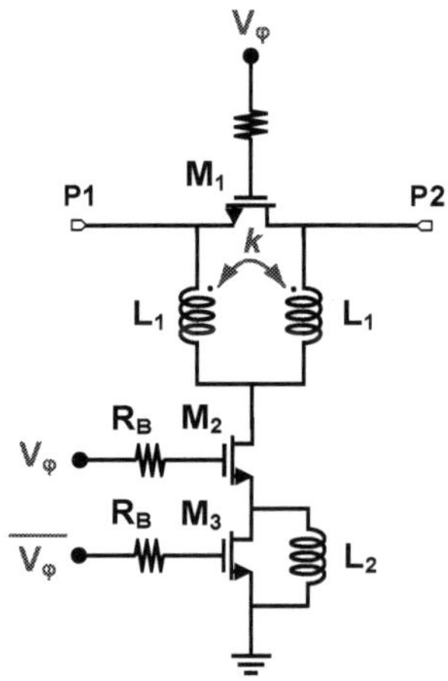

Figure 3.9 Topology of cross-coupled bridged T-type phase shifter.

To cope with the issues, the cross-coupled bridged T-type phase shifter is proposed for miniaturized phase shifter designs, which include mutual coupling effects and parasitic capacitance of transistor M_1 with careful designs considerations and modeling.

In Figure 3.9, the proposed cross-coupled bridged T-type phase shifter comprises two magnetic coupled inductors (L_1) with coupling coefficient k, another inductor (L_2), and three MOS transistors ($M_{1\text{-}3}$). The on-resistances of transistors are relatively small and ignored in the following analysis. Corresponding to the two operation states selected by control voltages (V_{φ}), the equivalent circuits of two states are given in Figure 3.10 with the relation as

$$L_M = k \times L_1 . \tag{3.12}$$

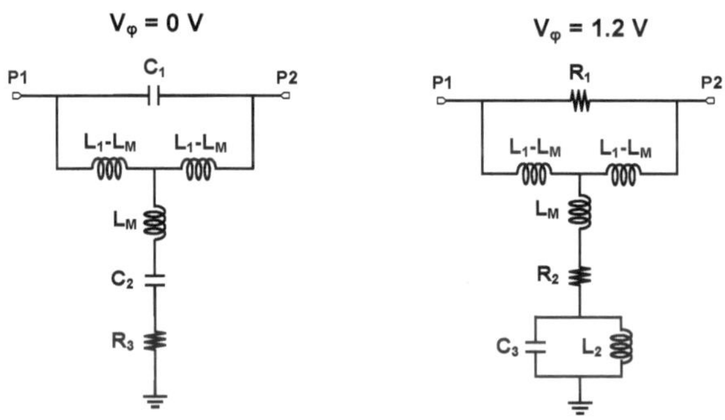

(a) (b)

Figure 3.10 Operation states of the cross-coupled bridged T-type phase shifter: (a) Vφ = 0 V; (b) Vφ = 1.2 V. (R_1, R_2 and R_3 denote the on-resistances of M1, M2 and M3; C_1, C_2 and C_3 denotes the off-capacitances of M1, M2 and M3)

When Vφ = 0 V, transistors M1 and M2 are off and transistor M3 is on, and the circuit is simplified to Figure 3.10(a). Since the equivalent circuit is symmetrical, reflection coefficient S_{11} and transmission coefficient S_{21} are derived in (3.13) and (3.14):

$$S_{11}=\frac{j\omega\left(\omega^2L_1C_2(1-k)\left(L_1(1+k)-2C_1Z_0^2\right)-2L_1(1-k)+C_2Z_0^2\right)}{\left(Z_0\left(1-2\omega^2L_1C_1(1-k)\right)+j\omega L_1(1-k)\right)\times\left(\omega^2L_1C_2(1+k)-j\omega C_2Z_0-2\right)} \tag{3.13}$$

$$S_{21}=\frac{-2Z_0\left(\omega^4L_1^2C_1C_2\left(1-k^2\right)-\omega^2L_1\left(2C_1(1-k)+kC_2\right)+1\right)}{\left(Z_0\left(1-2\omega^2L_1C_1(1-k)\right)+j\omega L_1(1-k)\right)\times\left(\omega^2L_1C_2(1+k)-j\omega C_2Z_0-2\right)} \tag{3.14}$$

Based on the transmission coefficient in (3.14), the phase shift and its derivative at frequency ω_0 are derived in (3.15) and (3.16):

$$\varphi_{0V}\big|_{\omega=\omega_0}=-\tan^{-1}\left(\frac{\omega_0L_1(1-k)}{Z_0\left(1-2\omega_0^2L_1C_1(1-k)\right)}\right)-\tan^{-1}\left(\frac{\omega_0C_2Z_0}{2-\omega_0^2L_1C_2(1+k)}\right), \tag{3.15}$$

$$\left.\frac{d\varphi_{0V}}{d\omega}\right|_{\omega=\omega_0}=\frac{2C_2\left(C_1C_2Z_0^4+k^2L_1^2-L_1^2\right)(1-k)\left(L_1+kL_1-2C_1Z_0^2\right)}{Z_0\begin{pmatrix}k^3L_1^2C_2+k^2\left(8L_1C_1^2Z_0^2-L_1^2C_2+L_1C_2^2Z_0^2+2L_1C_1C_2Z_0^2\right)\\+8L_1C_1^2Z_0^2+L_1^2C_2+k\begin{pmatrix}4L_1C_1C_2Z_0^2-L_1^2C_2-16L_1C_1^2Z_0^2\\+L_1C_2^2Z_0^2-C_1C_2^2Z_0^4\end{pmatrix}\\+C_1C_2^2Z_0^4-6L_1C_1C_2Z_0^2\end{pmatrix}} \tag{3.16}$$

To satisfy the impedance matching condition of $S_{11} = 0$ and phase shift of φ_0 at frequency ω_0, the circuit elements are solved using (3.13) and (3.15), and are expressed in (3.17) and (3.18) as follows:

$$L_1 = \frac{Z_0 \tan\left(\frac{\varphi_0}{2}\right)}{\omega_0 (1-k)\left(1+2\omega_0 C_1 Z_0 \tan\left(\frac{\varphi_0}{2}\right)\right)}, \tag{3.17}$$

$$C_2 = \frac{2\tan\left(\frac{\varphi_0}{2}\right)(1-k)\left(1+2\omega_0 C_1 Z_0 \tan\left(\frac{\varphi_0}{2}\right)\right)}{\omega_0 Z_0\left(\left(1+2\omega_0 C_1 Z_0 \tan\left(\frac{\varphi_0}{2}\right)\right)(1-k)+\tan^2\left(\frac{\varphi_0}{2}\right)(1+k)\right)}. \tag{3.18}$$

When $V_\varphi = 1.2$ V, transistors M_1 and M_2 are on and transistor M_3 is off, and the signal passes through the on-resistance of M_1 and M_2 while the off-capacitance of M_3 forms a band-pass filter structure with inductor L_2 as shown in Figure 3.10(b). Following the similar analysis steps with transistor on-resistance ignored, the following expressions for condition $S_{11} = 0$, phase shift, and phase derivative are derived at frequency ω_0:

$$L_2 = \frac{1}{\omega_0^2 C_3}, \tag{3.19}$$

$$\varphi_{1.2\mathrm{V}}\Big|_{\omega=\omega_0}=\tan^{-1}\left(\frac{Z_0}{\omega\left(L_1+\dfrac{2L_2}{1-\omega^2L_2C_3}\right)}\right)\Bigg|_{\omega=\omega_0}=0,\tag{3.20}$$

$$\left.\frac{d\varphi_{1.2\mathrm{V}}}{d\omega}\right|_{\omega=\omega_0}=-C_3Z_0.\tag{3.21}$$

Thus, phase shift φ_0 of the proposed phase shifter at frequency ω_0 is equal to $\varphi_{0\mathrm{V}}$ - $\varphi_{1.2\mathrm{V}}$ as in (3.22):

$$\varphi_0\Big|_{\omega=\omega_0}=-\tan^{-1}\left(\frac{\omega_0L_1(1-k)}{Z_0\left(1-2\omega_0^2L_1C_1(1-k)\right)}\right)-\tan^{-1}\left(\frac{\omega_0C_2Z_0}{2-\omega_0^2L_1C_2(1+k)}\right).\tag{3.22}$$

To obtain a broadband phase shift at frequency ω_0, it is necessary to satisfy the equality of first-order phase derivatives for the two states as:

$$\left.\frac{d\varphi_{0\mathrm{V}}}{d\omega}\right|_{\omega=\omega_0}=\left.\frac{d\varphi_{1.2\mathrm{V}}}{d\omega}\right|_{\omega=\omega_0}.\tag{3.23}$$

Thus, all the circuit parameters (L_1, L_2, C_1, C_2, C_3, k) are considered in the general design equations as in (3.17), (3.18), (3.19), (3.21), (3.22), and (3.23). Several numerical design examples based on above equations are solved to illustrate the phase responses and the magnetic coupling effects in cross-coupled bridged T-type phase shifters. Under the assumption that C_1 = 20 fF, Z_0 = 50 Ω, $\omega_0 = 2\pi \times 60$ GHz, and φ_0 = 90°, the circuit parameters are calculated for different k values using the above equations and summarized in Table VI.

Table VI Calculated design parameters for 90° designs.

k	L_1 (pH)	C_2 (fF)	L_2 (pH)	C_3 (fF)
0	75.6	67.6	57	123.5
0.05	79.6	65.1	55.6	126.6
0.1	84	62.5	54.4	129.2
0.15	89	59.9	51.8	135.9
0.2	94.5	57.2	51	138

The phase responses are plotted in Figure 3.11. In all the cases, the exact 90° phase shifts are obtains at frequency ω_0, while phase responses at frequency ω_0 are flat as expected from (3.23). For different coupling strengths, the design with zero coupling has the lowest phase deviation from 90° across frequency, however requires large space between the two inductors in silicon implementations. In fact, phase shifters with $k < 0.15$ still provide good phase responses with phase error < 5° from 50 to 70

GHz. Thus, it is more favorable to reduce the inductors spacing for compact circuit sizes, while still obtain acceptable phase accuracy by co-designing with the presence of cross-coupling effects.

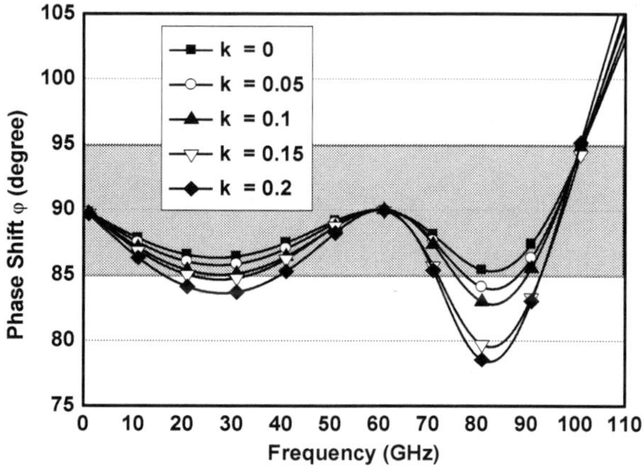

Figure 3.11 Ideal phase responses of the cross-coupled bridged T-type phase shifter for C_1 = 20 fF, Z_0 = 50 Ω, $\omega_0 = 2\pi \times 60$ GHz, φ_0 = 90°, and k = 0, 0.05, 0.1, 0.15, 0.2.

3.3.2 60-GHz 0/180° and 0/90° Phase Shifters

The design is based on GLOBALFOUNDRIES (GF) 65-nm CMOS technology. The 3-D full-wave EM field simulator ANSYS HFSS v.14 is used for simulations of the inductors, coupling coefficient, interconnects, vias, and testing pads. Cadence is used for circuit co-simulations with foundry provided process design kits (PDK).

The 0/180° phase shifter is designed by cascading two 0/90° phase shifters. Inductors are implemented using transmission lines (TL), where the 3.3-μm thick M8 layer is used as signal line and the 0.22-μm thick M1 forms the ground plane as shown in Figure 3.12. Metal stack information is depicted in Figure 3.1, which also provides design parameters that are optimized with circuit co-simulations.

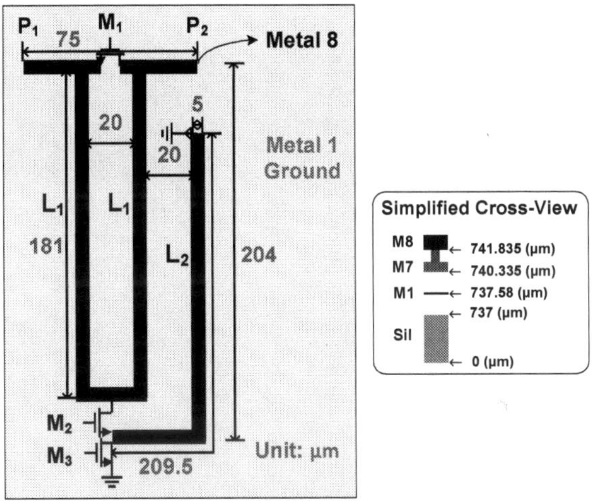

Figure 3.12 Configuration of the 60-GHz 0/90° phase shifter.

Based on Figure 3.11, coupling coefficient $k = 0.05$ is chosen as the optimum tradeoff between phase performance and circuit size. With $k = 0.05$, the coupled inductors L_1 can be placed as close as 20 μm only from EM simulations. Compared to the conventional 60-GHz 0/90° T-type phase shifter in [13], this prototype achieves size reduction of more than 65%. In Figure 3.13, the simulated results of phase shifter have insertion loss of ~2.5 dB and phase shift of 90±2.5° from 57 to 64 GHz.

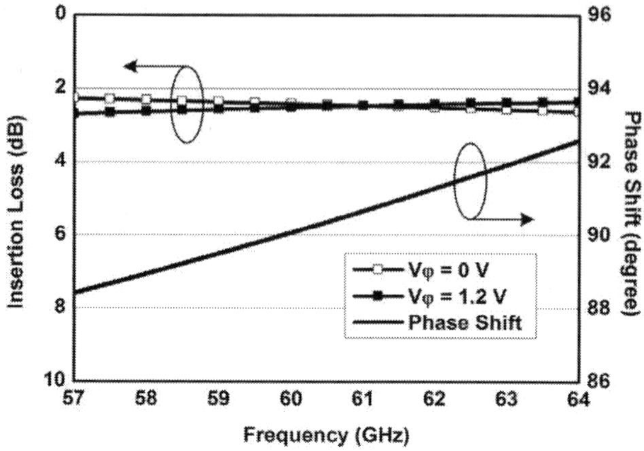

Figure 3.13 Simulated performance of the 60-GHz 0/90° phase shifter.

3.4 The 60-GHz 5-bit 360° Phase Shifter

3.4.1 Architecture

As in Figure 3.14, the 60-GHz 5-bit phase shifter is composed of the cross-coupled bridged T-type STPSs for 2-bit coarse phase control (0/180°, 0/90°) and a digital RTPS for 3-bit fine phase control (0/45°, 0/22.5°, 0/11.25°). Therefore, the completed phase shifter covers a full 360° phase shift with a fine 5-bit digital phase control. The 3-bit 90° phase shifter has been discussed in section 3.2, while the 0/180° and 0/90° phase shifters were described in Chapter 3.3.

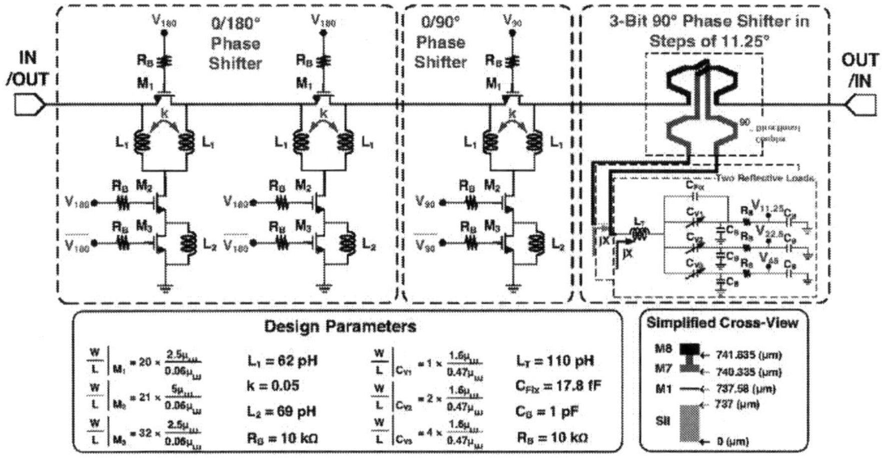

Figure 3.14 Schematic of the 5-bit phase shifter design with design parameters and simplified cross-view of the used process.

Compared to the conventional 360° STPSs in [11]-[17], this architecture adopts the cross-coupled bridged T-type STPS that provides accurate phase shifts with a very compact circuit size. Besides, it also emulates the STPSs of 0/45°, 0/22.5°, and 0/11.25° phase shifts with switch-varactor RTPS that has the similar 3-bit phase control function, but achieves much more compact size and lower insertion loss.

3.4.2 Implementation

The design is based on GLOBALFOUNDRIES (GF) 65-nm CMOS technology. The phase shifters are designed with 50-Ω input/output matching and cascaded in series to obtain 5-bit phase control. The 3-D full-wave EM field simulator ANSYS HFSS v.14 is used for simulations of the inductors, coupling coefficient, interconnects, vias, and testing pads. Cadence is used for circuit co-simulations with foundry provided process design kits (PDK).

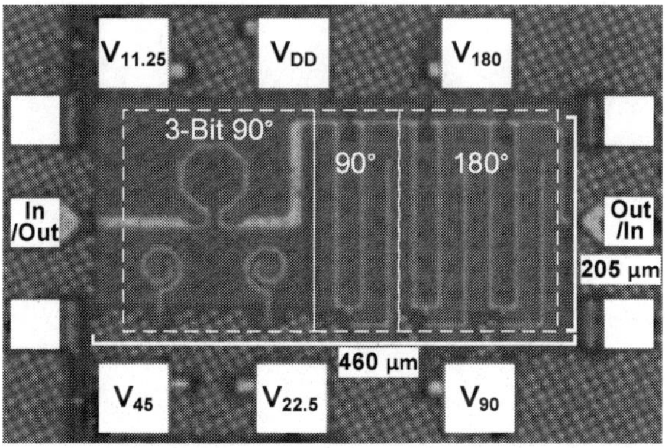

Figure 3.15 Micrograph of the 60-GHz 5-bit 360° phase shifter chip die.

In Figure 3.15, the fabricated 60-GHz 5-bit phase shifter occupies a size of 0.094 mm^2 only (excluding pads). The power supply V_{DD} of 1.2 V is used for two on-chip inverters that simplify the control logic of the 0/180° and 0/90° phase shifters. The phase shifter consumes less than 1 nA current that is mainly due to transistor leakages.

3.4.3 Measurements

Two-port S-parameter measurements are performed from 1 GHz to 110 GHz using an Agilent 67GHz PNA-X, a OML's extension module and Cascade i-110 probes. The system is calibrated using a SOLT probe-tip calibration using a Cascade ISS substrate. Therefore, the RF pad loss is included in the insertion loss measurements.

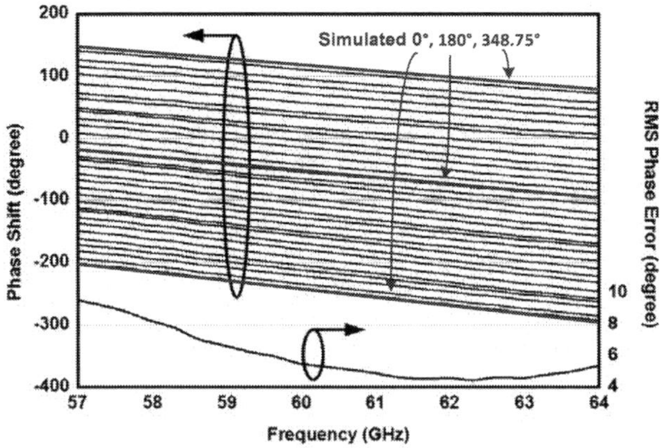

Figure 3.16 Measured phase responses for 32 states and RMS phase error of the 60-GHz 5-bit 360° phase shifter.

In Figure 3.16, the measured phase responses for 32 states are plotted. The phase shifter demonstrates 5-bit phase control covering 360° from 57 to 64 GHz. The measured RMS phase error is 4.4-9.5° in the operating bandwidth, indicating good phase linearity. The maximum phase error in a single state is 7.7° at 60 GHz, 11.8° at 57 GHz, and 7.8° at 64 GHz.

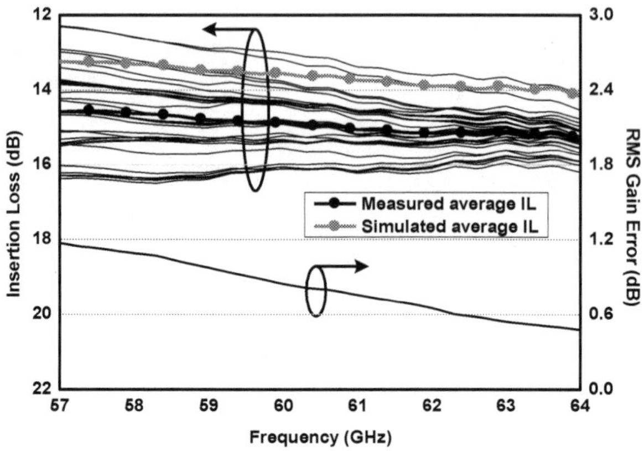

Figure 3.17 Measured insertion loss for 32 states and RMS gain error of the 60-GHz 5-bit 360° phase shifter.

In Figure 3.17, the simulated and measured average insertion losses for 32 states are in good agreement, where the measured average is ~1.5 dB larger due to the simplifications in EM simulations of metal vias and testing pads. The measured average insertion loss is 14.3 dB from 57 to 64 GHz with a variation of ±2 dB. The small insertion loss variation of 2 dB relaxes the design constraints and complexity of the RF variable-gain amplifiers, which are used in beam-forming front-ends to compensate the insertion loss variation of phase shifters [3, 5, 16]. As a 5-bit phase shifter, the design has insertion loss per control bit of only 2.86 dB. The RMS gain error calculation uses the definition in [71] as

$$\text{RMS gain error} = \sqrt{\frac{1}{\mathrm{N}} \times \sum_{i=1}^{\mathrm{N}} \left| \mathrm{IL}_i \text{-} \mathrm{IL}_{\text{average}} \right|^2}, \tag{3.24}$$

which is calculated as 0.5-1.1 dB from 57 to 64 GHz. In Figure 3.18, the measured input and output return loss are better than 10 dB in all states from 57 to 64 GHz.

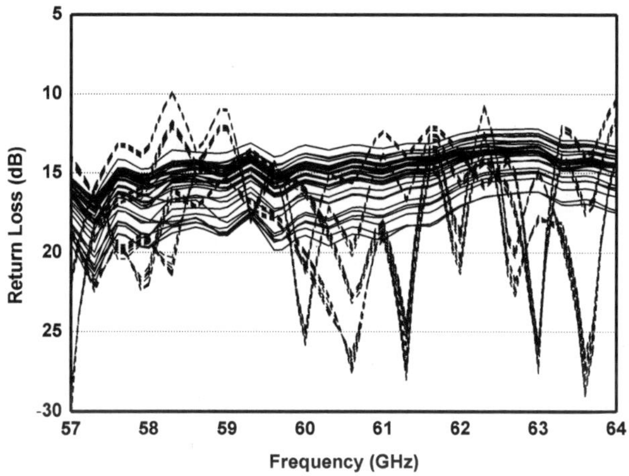

Figure 3.18 Measured return loss for 32 states of the 60-GHz 5-bit 360° phase shifter. (Dash lines: input return loss; Solid lines: output return loss)

The power performance of the phase shift is measured using an Agilent 67GHz PNA-X, an R&S ZVA Vector Network Analyzer used as frequency generators, a Cernex V-band power amplifier, and Cascade probes. Figure 3.19 shows the output power for 32 states versus input power. The measured input P_{1dB} is better than 9.5 dBm, which is enough for beam-forming applications with medium-large delivery power [51].

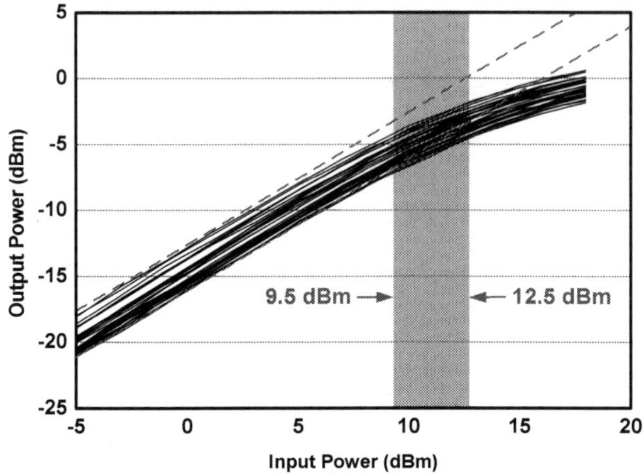

Figure 3.19 Measured power performance for 32 states of the 60-GHz 5-bit 360° phase shifter.

The performance summary of the proposed 60-GHz 5-bit 360° phase shifter and a comparison with other similar phase shifters is shown in Table VII. Compared to the state-of-the-arts, the proposed phase shifter achieves the lowest insertion loss per bit, the most compact size, and small RMS phase/gain errors.

Table VII Performance summary and comparison of state-ofthe-art mm-wave 360° phase shifters.

Ref.	Process	Frequency (GHz)	IL_{MAX} (dB)	Total Phase Shift (°)	RMS Gain Error (dB)	RMS Phase Error (°)	Core Area (mm²)	P_{1dB} (dBm)	FoM ** (°/dB)
[86] MWCL 2009	CMOS 90-nm	50-65	8	90	*	*	0.08	4	11.25
[87] RFIC 2009	BiCMOS 0.13-μm	57-64	8	180	0.2	2.7	0.18	*	22.5
[12] MWCL 2014	CMOS 65-nm	54-66	6.9	90	*	5.2	0.034	*	13
[51] TCAS-	CMOS 65-nm	75-85	27.1	360	1-1.8	7.2-11.2	0.122	>15	13.3

II 2014									
[64] IMS 2012	CMOS 90-nm	57-64	15.6	360	1.3	*	0.28	*	23.1
[90] CISC 2009	BiCMOS 0.12-μm	67-78	30	360	1.5-2.8	3.5-12	0.135	>8	12
[13] TMTT 2013	CMOS 90-nm	57-64	18	360	1.6-1.8	2-10	0.34	*	20
This Work	CMOS 65-nm	57-64	16.3	360	0.5-1.1	4.4-9.5	0.094	9.5-12.5	22.1

* Not mentioned # Pad loss included ** Total phase shift / Maximum insertion loss

† VM: vector modulation; STPS: switched-type phase shifter; RTPS: reflective-type phase shifter

3.5 Summary

The chapter presents two types of novel phase shifter topologies, namely the switched-varactor type phase shifter and the cross-coupled bridged T-type phase shifter. By co-designing phase shifters using the proposed topologies, a 60-GHz 5-bit 360° phase shifter is designed, implemented, and successfully verified experimentally. Low-loss and low phase/gain-errors are achieved. Further, the miniaturized design could have high potential in phase-array applications, especially in large-scale arrays.

CHAPTER 4

MM-wave Bidirectional Beam-forming Front-End

4.1 Introduction

Recently, intensive research efforts have been focused on the time division duplex (TDD) transceivers leveraging on bidirectional techniques for the IEEE 802.11ad standard commercial applications [3, 143]. In these applications where power and cost are majorly concerned, the associated building blocks are required to have low power consumption, compact circuit size, and low fabrication cost.

The conventional bidirectional transceiver uses single-pole double-throw (SPDT) switches for amplifier selection and isolation [143]. However, each SPDT switches has insertion loss of about 2-4 dB at 60 GHz [11, 144]. This insertion loss directly adds to system noise figure (NF) in the Rx path, while reduces output power (P_{OUT}) in the Tx path. To cope with the issue, we propose a new architecture of a four-element bidirectional beam-former as shown in Figure 4.1. Each element comprises a 60-GHz bidirectional low-noise amplifier power amplifier (LNAPA), and a 5-bit phase shifter that is presented in Chapter 3. Then, Wilkinson power combiners are used as 4-to-1 power combiner in receive mode and 1-to-4 power splitter in transmit mode.

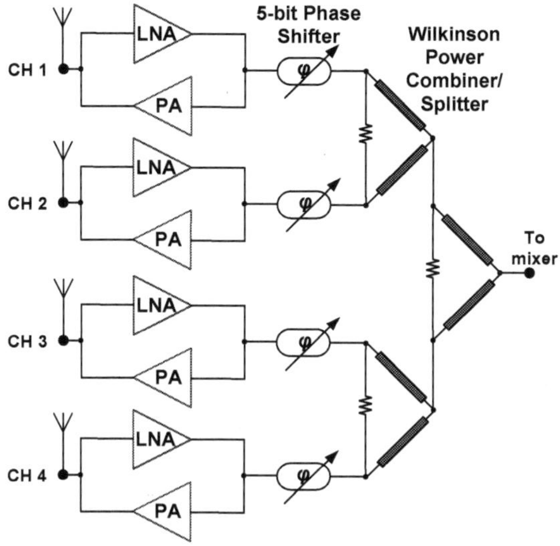

Figure 4.1 System architecture of the proposed bidirectional 4-element beam-former.

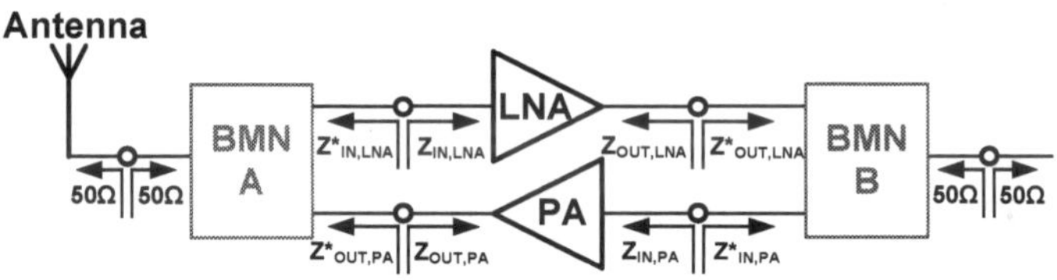

Figure 4.2 Diagram of the proposed bidirectional LNAPA using *BMN*s.

In this chapter, the bidirectional LNAPA is presented firstly. The design alleviates the lossy RF switches used in conventional bidirectional amplifier designs, the proposed three-port bidirectional matching networks (*BMN*s) (as shown in Figure 4.2) to provide input/output impedance matching for optimum NF and output power of LNA and PA core circuits, respectively. To validate the concept, a 60 GHz LNAPA prototype is designed, fabricated, and verified experimentally in a 65-nm CMOS technology. Furthermore, the entire 60-GHz 4-element beam-former is designed and fabricated. Due to the time constraint, the beam-former has not been measured and published. Thus, only circuit layout and preliminary simulation results are provided in the book.

4.2 Bidirectional LNAPA

4.2.1 Bidirectional Matching Network

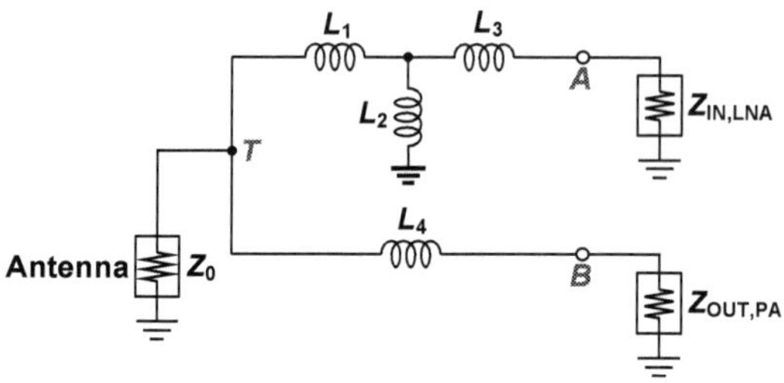

Figure 4.3 Schematic of the bidirectional matching network.

Figure 4.3 shows the proposed *BMN* at the antenna port, which comprises four inductive components $L_{1\text{-}4}$. Based on the results from noise-contour simulation of the LNA core input and load-pull simulation of the PA core output, the optimum LNA input impedance $Z_{IN,LNA}$ and optimum PA output impedance $Z_{OUT,PA}$ can be obtained.

Then, the four matching components $L_{1\text{-}4}$ can provide conjugate matching at points A, B, T using (1)-(3) as follows:

$$Z_{\mathrm{IN,LNA}}{}^* = \left(\left(Z_{\mathrm{OUT,PA}} + sL_4\right) // Z_0 + sL_1\right) // sL_2 + sL_3, \tag{4.1}$$

$$Z_{\mathrm{OUT,PA}}{}^* = \left(\left(Z_{\mathrm{IN,LNA}} + sL_3\right) // sL_2 + sL_1\right) // Z_0 + sL_4, \tag{4.2}$$

$$Z_0 = \left(\left(Z_{\mathrm{IN,LNA}} + sL_3\right) // sL_2 + sL_1\right) // \left(Z_{\mathrm{OUT,PA}} + sL_4\right). \tag{4.3}$$

According to the proposed topology and theoretical analysis, with consideration of the finite Q-factors of $L_{1\text{-}4}$, the optimum design can be achieved by setting smaller inductance values. In this work, we use *BMN* circuit co-designed with LNA and PA core circuits, and achieve the optimized tradeoff between NF and P_{OUT} of the designed bidirectional LNAPA.

4.2.2 Bidirectional LNAPA Design

Figure 4.4 shows the schematic of proposed LNAPA, comprising of LNA core, PA core, and two *BMN*s. Both networks are designed based on (4.1)-(4.3) in our previous analysis. It is noted that the *BMN* B is also matched to 50 Ω output impedance for measurement purpose.

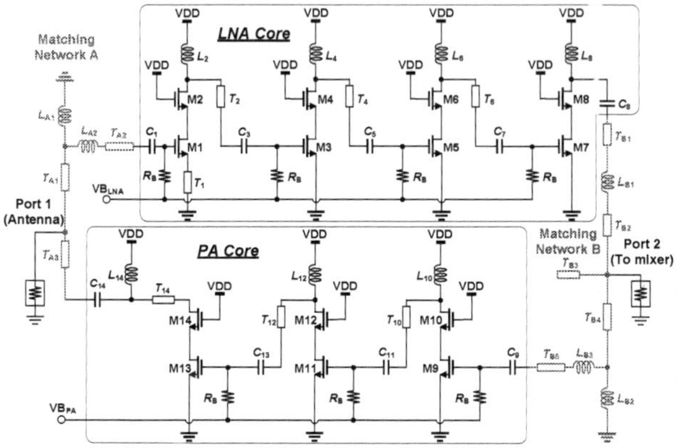

Transistor	Width(μm)/ Length(μm)
M1	22/0.06
M2	22/0.06
M3	32/0.06
M4	32/0.06
M5	32/0.06
M6	32/0.06
M7	32/0.06
M8	32/0.06
M9	36/0.06
M10	36/0.06
M11	48/0.06
M12	48/0.06
M13	60/0.06
M14	60/0.06

Inductor	(pH)
L_{A1}	60
L_{A2}	64
L_{B1}	106
L_{B2}	53
L_{B3}	57
L_2	113
L_4	81
L_6	75
L_8	120
L_{10}	106
L_{12}	57
L_{14}	50

Transmission Line	(μm)
T_{A1}	70
T_{A2}	70
T_{A3}	130
T_{B1}	30
T_{B2}	140
T_{B3}	340
T_{B4}	120
T_{B5}	70
T_1	40
T_2	40
T_4	50
T_6	50
T_{10}	50
T_{12}	50
T_{14}	160

Capacitor	(fF)
C_1	188
C_3	69
C_5	46
C_7	68
C_8	57
C_9	103
C_{11}	202
C_{13}	74
C_{14}	129

Resistor	kΩ
R_B	8

Figure 4.4 Schematic and design parameters of the proposed bidirectional LNAPA. (*BMN* A and B are marked in red and blue colors)

In receive (Rx) mode, the LNA core circuit is active with nominal gate bias voltage VB_{LNA} while PA core circuit is inactive with VB_{PA} of 0 V. The LNA core is designed to provide small-signal gain of 20 dB, bandwidth of 57-66 GHz, and NF of 6 dB. To achieve these targets, the LNA core adopts four-stage cascode topology with on-chip spiral inductor loads ($L_{2,4,6,8}$). Short length TLs ($T_{2,4,6}$) are used for inter-stage connections, while capacitors ($C_{1,3,5,7,8}$) provide blocking for gate biasing. Large resistors (R_B) are effectively open-circuit for RF signals and used for gate biasing. For low noise operation, transistors *M1* and *M2* are selected with width of 22 μm and biased at low current density of 0.175 mA/μm.

In transmit (Rx) mode, the PA core circuit is active with nominal gate bias voltage VB_{PA} while LNA core circuit is inactive with VB_{LNA} of 0 V. The PA core is designed to provide small-signal gain of 20 dB, bandwidth of 57-66 GHz, P_{SAT} of 10 dBm, and PAE of 10%. The PA core uses three-stage cascode topology. Similar to LNA core, its associated inductive loads, inter-stage connections, DC blockings, and gate biasing are achieved by using inductors ($L_{12,14,16}$), TLs ($T_{10,12,14}$), capacitors ($C_{9,11,12,14}$), and resistors (R_B) respectively. To obtain output power of 10 dBm, transistors *M13* and *M14* are chosen with width of 60 μm and operate in Class-A mode. The prior stages are designed for linear amplification with enough power to drive the following stages.

4.2.2 Measurement

The bidirectional LNAPA is fabricated in GLOBALFOUNDRIES (GF) 65-nm CMOS technology (f_T = 220 GHz). The chip micrograph is shown in Figure 4.5, where the circuit occupies a size of 0.48 mm × 0.46 mm (0.22 mm^2). At nominal bias conditions, the circuit consumes 39.6 mW and 71.1 mW in Rx and Tx modes respectively with 1.8 V supply.

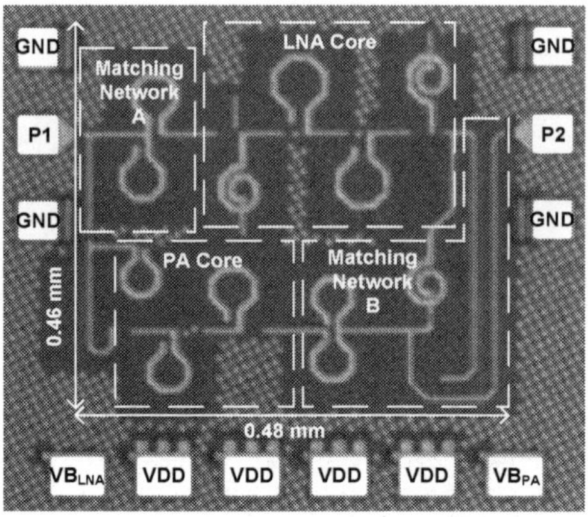

Figure 4.5 Micrograph of fabricated bidirectional LNAPA.

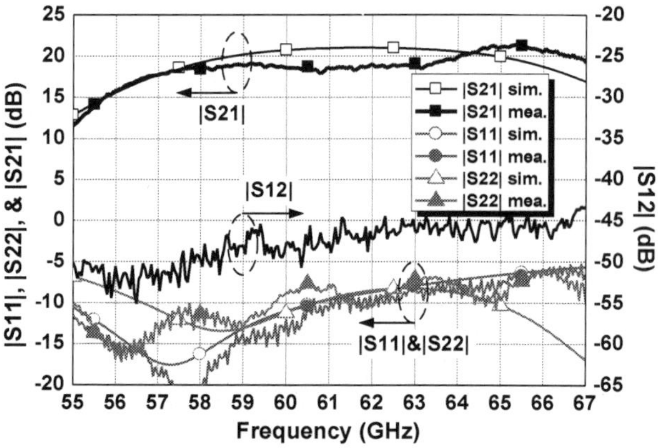

Figure 4.6 S-parameters in the Rx mode.

Small-signal measurements are performed from 55-67 GHz using an Agilent 67-GHz PNA-X and Cascade probes. As shown in Figure 4.6, in the Rx mode with VB_{LNA} = 0.8 V and VB_{LNA} = 0 V, the amplifier has measured peak gain of 21.5 dB at 65.5 GHz and gain of >17 dB over 57-67 GHz. Return losses are better than 7 dB, while isolation is better than 43 dB. Figure 4.7 shows the PA mode performance with $VB_{LNA} = 0$ V and $VB_{LNA} = 1.8$ V. The amplifier has measured peak gain of 24.5 dB at 59 GHz and gain of >17 dB over 57-65 GHz. Return losses are better than 8 dB, while isolation is better than 43 dB.

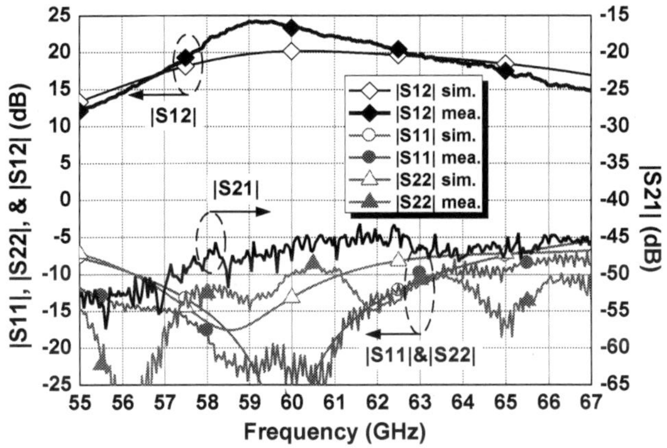

Figure 4.7 S-parameters in the Tx mode.

Noise performance in Rx mode is characterized using a PNA-X, a power sensor, and a Cernex V-band PA. In Figure 4.7, the measured minimum NF is 6.7 dB at 58.5 GHz. The simulated NF of proposed amplifier and its LNA core with typical matching networks are compared. It is noted that the *BMN* degrades NF by ~1.4 dB that is much lower than the typical insertion loss of SPDT switches at 60 GHz.

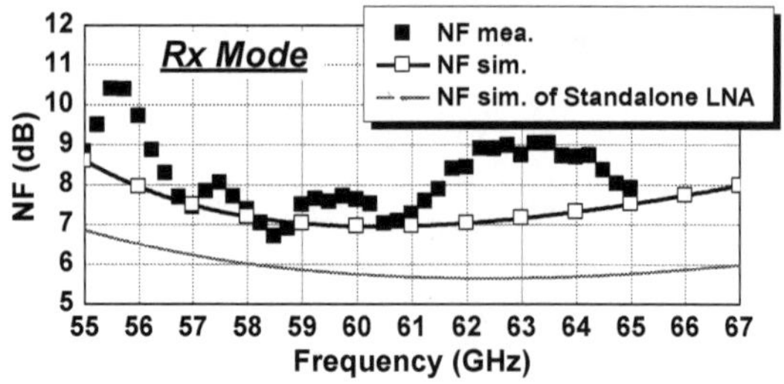

Figure 4.8 NF in the Rx mode.

Power performance in Tx mode is evaluated using a PNA-X and an R&S VNA as frequency generator at 60 GHz. In Figure 4.9, the amplifier achieves output P_{1dB} of 5 dBm, P_{SAT} of 8.4 dBm with PAE of 8.7% when P_{IN} is -3 dBm. It agrees well with the simulated P_{1dB} of 5.3-5.5 dBm.

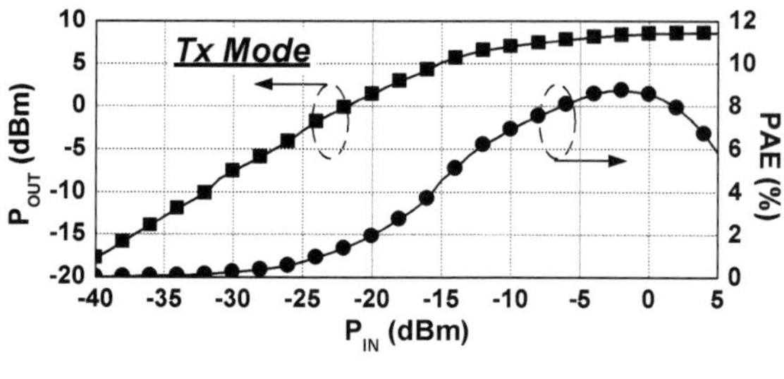

Figure 4.9 Power performance in the Tx mode.

Table VIII shows the performance summary of proposed bidirectional LNAPA and the comparison with bidirectional amplifier using SPDT switches [143], standalone LNA [145, 146] and PA [147, 148]. In Rx mode, the proposed amplifier achieves high gain and comparable NF. In Tx mode, high gain and good P_{SAT}/PAE are obtained with low power consumption. It is noted that the proposed amplifier is very compact in size and suitable for low-cost transceivers, especially in large-scale arrays.

Table VIII Performance Summary and Comparison of State-of-the-Arts.

Ref.	Tech.	Freq. (GHz)	Gain (dB)	NF (dB)	Return Loss (dB)	Size (mm^2)	Supply Voltage (V)	P_{DC} (mW)	FoM †
[149] Bi. Amp (forward)	75-nm InP	103-153	15	5**	5	0.1	2.4	15	22.2
[145] LNA only	65-nm CMOS	56-61	16.5	6.6	15	0.37	1.5	27.9	3.9
[146] LNA only	90-nm CMOS	58-65	11.3	4.4	>10	0.27	2.4	14.4	7
[143] LNA & Switch	65-nm CMOS	54-66	17.8	5.6	>10	1*	1.0	13.2	12.2
This work Rx mode	65-nm CMOS	57-67	21.5	6.7	>7	0.22#	1.8	39.6	4.8
Ref.	**Tech.**	**Freq. (GHz)**	**Gain (dB)**	**P_{SAT} (dBm)**	**Return Loss (dB)**	**PAE (%)**	**Supply Voltage (V)**	**Size (mm^2)**	**P_{DC} (mW)**
[149] Bi. Amp (reverse)	75-nm InP	70-110	12	5	5	-	2.4	0.1	15
[147] PA only	65-nm CMOS	54-66	17.7	16.8	6	14.5	1.5	0.32	378
[148]PA only	65-nm CMOS	44-60	8.3	11	10	7.1	1.2	0.39	130
[143] PA & Switch	65-nm CMOS	54-64	21.5	5.6	>7.6	3.2	1.0	1*	96.2
This work Tx mode	65-nm CMOS	57-65	24.5	8.4	>9	8.7	1.8	0.22 #	71.1

* Includes SPDT switches, LNA, and PA ** Simulation

Total size of bidirectional LNAPA

† $FoM[\text{GHz/mW}] = \dfrac{Gain\,[\text{abs}] \times f_C[\text{GHz}]}{(F-1)\times P_{DC}[\text{mW}]}$

4.3 Power Combiner/ Splitter

In the proposed beam-former, a power handling block which is used as 4-to-1 power combiner in the receiving mode and as a 1-to-4 power splitter in the transmitting mode has to be developed in CMOS.

In this work, passive Wilkinson power combiner/splitter topology is adopted. Three 50-Ω matched 2-to-1 power combiners are arranged as in Figure 4.1 and optimized in EM simulation with HFSS as shown in Figure 4.10.

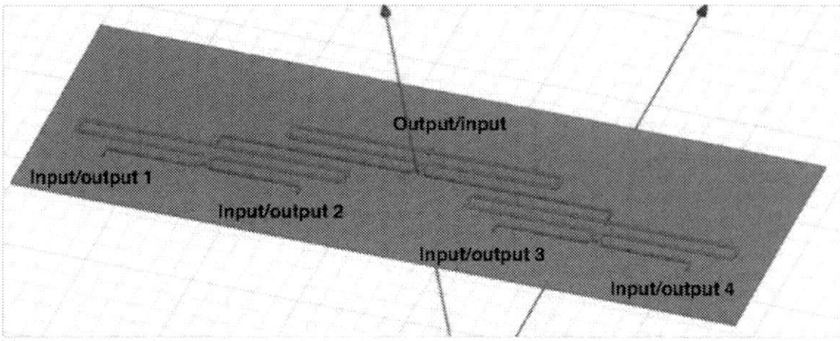

Figure 4.10 Power combiner/splitter in EM simulation environment.

The EM simulation results are plotted in Figure 4.11. The total insertion loss including the 6-dB power splitting loss is 7.3 dB from 50 to 70 GHz. The input/output matching conditions are better than 15 dB. In-band port-to-port isolations are better than 20 dB.

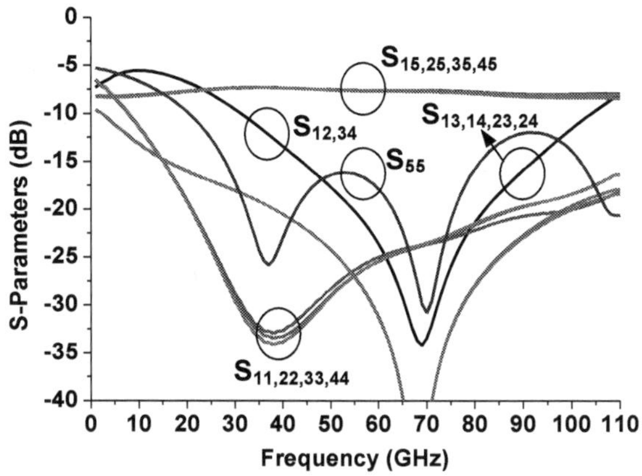

Figure 4.11 Simulated S-parameters of the power combiner/splitter.

4.4 60-GHz 4-Element Bidirectional Beam-Former

The entire 60-GHz 4-element bidirectional beam-former was completed as shown in Figure 4.12. The fabricated chip has returned to our group. However, due to the fact that the beam-former measurement requires 4-channel simultaneous 60-GHz input to measure its beam-forming capability, it requires packaging LTCC to perform the testing as shown in Figure 4.13. It has not been completed yet.

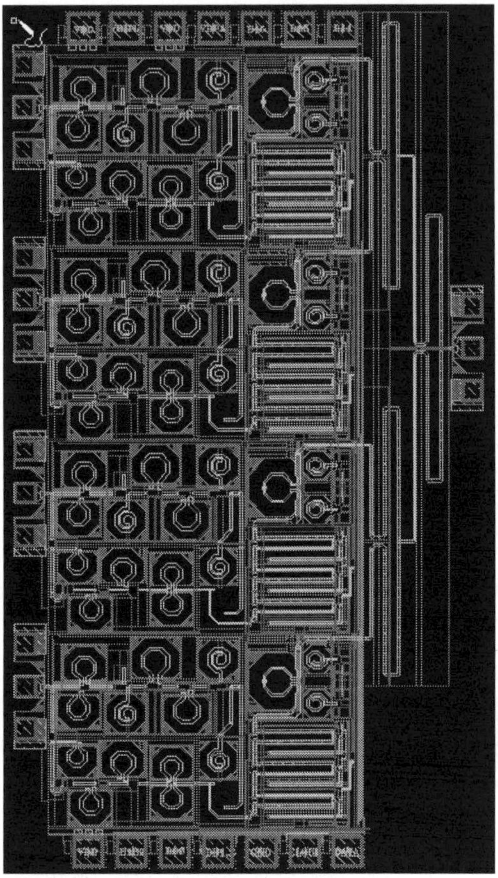

Figure 4.12 Cadence layout of the proposed bidirectional 4-element beam-former (2mm × 1mm).

However, individual key building blocks and their measured results have been performed and reported in previous chapters. A good beam-forming capability is expected from the designed beam-former.

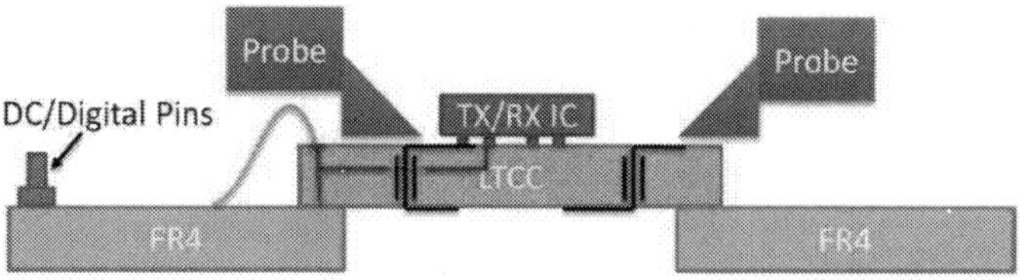

Figure 4.13 Experimental setup for array pattern measurement.

4.5 Summary

This chapter showed an mm-wave bidirectional LNAPA design without utilizing RF switches. High gain, good linearity, unconditionally stable, low power consumption, and compact size were demonstrated in the fabricated prototype. It should be noted that this work has high potential in commercial mm-wave transceivers and imaging systems, where simple designs with small form-factor and low power consumption are preferable. Furthermore, a full 4-element bidirectional beam-forming front-end was designed and fabricated. The chip will be measured with LTCC packages in the future.

CHAPTER 5

Conclusion and Future Prospects

5.1 Conclusion

This book focused primarily on the mm-wave IC designs in CMOS technology. Several key circuit blocks were proposed and analyzed, such as phase shifters, SPDT switches and bidirectional LNAPA. The proposed designs were fabricated and verified through on-chip measurements and competitive results were obtained.

Firstly, to cater for terahertz communication and imaging system applications, several mm-wave single-pole double-throw (SPDT) switches in 65-nm CMOS were demonstrated. A new topology, namely the magnetically switchable artificial resonator was proposed to alleviate the bulky and lossy transmission lines used in conventional SPDT designs. Three switch prototypes were designed and fabricated in a 65-nm CMOS technology; two operate at 130-180 GHz and one operates at 220-285 GHz. The fabricated 130-180 GHz SPDT switch achieves insertion loss of 3.3 dB, and isolation of 23.7 dB. To the authors' best knowledge, the circuit size of 0.0035 mm^2 is the smallest among SPDT switches with similar operating frequencies. The fabricated 220-285 GHz SPDT switch features measured insertion loss of 4.2 dB including RF pad losses, isolation of 19 dB, return loss of better than 10 dB, simulated P_{1dB} of 9.2 dBm, and zero power consumption. To the best of authors' knowledge, this switch achieves the highest operating frequency and smallest chip size among reported SPDT switches in CMOS and BiCMOS technologies.

Secondly, a compact 3-bit 90° phase shifter for phased-array applications at the 60-GHz ISM band (IEEE 802.11ad standard) was presented. The designed phase shifter is based on reflective-type topology using the proposed reflective loads with binary-weighted digitally-controlled varactor arrays and the transformer-type directional coupler. The measured eight output states of the implemented phase shifter in 65-nm CMOS technology exhibit phase-resolution of 11.25° with an RMS phase error of 5.2°. The insertion loss is 5.69±1.22 dB at 60-GHz and the return loss is better than 12 dB over 54-66 GHz. The chip demonstrates a compact size of only 0.034 mm^2.

Thirdly, a compact 60-GHz phase shifter that provides 5-bit digital phase control and 360° phase range for beam-forming systems was presented by co-designing the abovementioned 3-bit 90° phase shifter, 0/180° and 0/90° phase shifters. The 0/180° and 0/90° phase shifters were designed by using the proposed cross-coupled bridged T-type topology. The topologies are analyzed using small-signal equivalent circuit model. Further, the design equations are derived and investigated. To validate the theoretical analysis, the 60-GHz 5-bit 360° phase shifters was designed in a commercial 65-nm CMOS technology. The fabricated 360° phase shifter features good performance of 32 phase states from 57 to 64 GHz with RMS phase error of 4.4°, total insertion loss of 14.3±2 dB, RMS gain error of 0.5 dB, P_{1dB} of better than

9.5 dBm, and power consumption of almost zero. The average insertion loss is only 2.8 dB per control bit. To authors' best knowledge, the designed 360° phase shifter with the size of 0.094 mm^2 is the smallest 5-bit passive phase shifter at frequencies around 60 GHz.

At last, to alleviate RF switches in the promising bidirectional technique, a 56-67 GHz bidirectional LNAPA design was reported. The bidirectional matching networks were introduced to connect LNA and PA core circuits in parallel and satisfy isolation requirements with full consideration of input/output impedance matching of the LNA and PA. Thus, the operation modes are simply selected by alternated gate biasing of the LNA and PA core circuits. The fabricated amplifier achieves excellent performance; in the Rx mode, it features peak gain of 21.5 dB with gain of >17 dB over 56-67 GHz, NF of 6.7 dB with 39.6 mW power consumption, while in the Tx mode it achieves peak gain of 24.5 dB with gain of >17 dB over 56-65 GHz, P_{SAT} of 8.4 dBm, PAE of 8.7% with 71.1 mW power consumption. The reverse isolation in both modes is better than 43 dB. The circuit occupies a compact size of 0.22 mm^2.

5.2 Future Prospects

In this book, a complete 60-GHz beam-forming front-end with 5-bit phase tuning capability has been demonstrated. Thus, to design a complete beam-forming system including the present RF front-end with mixers, VGAs, DAC/ADC, and DSP blocks would be the most challenging and most interesting work in future. Such system operates in TDD mode without using any RF switches, and achieves low power consumption as well as low form-factor. It is very suitable for the next-generation WLAN applications, especially in the commercial portable and mobile devices.

In the past few years, some pioneer works have demonstrated THz frequency generations using novel architectures and passive building blocks in CMOS technology. These works are primarily done by Prof. Afshari's research group in Cornell University and Prof. Han's research group in Massachusetts Institute of Technology. It could save tremendous financial cost to the existing THz systems, if such frequency generators are designed and optimized properly to meet the commercialization requirements. In this research work, some novel design techniques used in the CMOS SPDT switches, namely the coupled-line structures and magnetic switchable concept, could be extended and exploited in the frequency generator designs to further reduce the passive loss and enhance isolations.

Current mm-wave researches from both academy and industries have showed an inevitable trend towards much larger scale of beam-former developments. The main reason is to cater for the non-light-of-sight (NLOS) and the long-distance communications, since mm-wave CMOS PA has showed very little improvements in terms of output power in the past few years. Moreover, current quartz wafer technology provides a suitable solution for the antenna developments for such large-scale array. It could be very interesting and challenging to develop a full-wafer array with beam-forming elements up-to tens of thousands. Thus, accurate, high-power, and highly-functional mm-wave transceivers or imagers could be realized for the next-generation applications.

Lastly, the author believes that mm-wave in-band full-duplex (IBFD) will be one of the hottest topics in the next decades. IBFD potentially allows system to maximally double the spectrum usage. However, the main challenge is on how to cancel the self-interference from the Tx to the Rx in the same device, especially in the condition with such wide operation bandwidth. Current works by Stanford University have demonstrated RF SI cancellation in a 20 MHz bandwidth. To extend it to several GHz at mm-wave is still challenging. The key technique is on how to maintain good wideband phase linearity and group delay. One possible solution is to use left-handed structure that has nearly opposite EM propagation property as compared to conventional transmission lines, and to accurately monitor the signal phase and group delay in the RF SI cancellation circuits.

BIBLIOGRAPHY

[1] K. Ma, *et al.*, "An integrated 60GHz low power two-chip wireless system based on IEEE802.11ad standard," in *IEEE MTT-S Int. Microw. Symp. Dig.*, 2014, pp. 1-4.

[2] K. Ma, S. Mou, Y. Wang, J. Yan, and K. S. Yeo, "A miniaturized 28mW 60GHz differential quadrature sub-harmonic QPSK modulator in 0.18um SiGe BiCMOS," in *IEEE MTT-S Int. Microw. Symp. Dig.*, 2014, pp. 344-345.

[3] M. Boers, *et al.*, "A 16TX/16RX 60GHz 802.11ad chipset with single coaxial interface and polarization diversity," in *IEEE Int. Solid-State Circuits Conf. (ISSCC) Dig.*, 2014, pp. 344-345.

[4] X. Yi, C. C. Boon, H. Liu, J. Lin, and W. M. Lim, "A 57.9-to-68.3 GHz 24.6 mW frequency synthesizer with in-phase injection-coupled QVCO in 65 nm CMOS technology," *IEEE J. Solid-State Circuits,* vol. 49, pp. 347-359, 2014.

[5] A. Natarajan, *et al.*, "A fully-integrated 16-element phased-array receiver in SiGe BiCMOS for 60-GHz communications," *IEEE J. Solid-State Circuits,* vol. 46, pp. 1059-1075, 2011.

[6] F. Meng, K. Ma, K. S. Yeo, S. Xu, C. C. Boon, and W. M. Lim, "A 60-GHz 26.3-dB gain 5.3-dB NF low-noise amplifier in 65-nm CMOS using Q-factor enhanced inductors " in *URSI General Assembly and Scientific Symp.*, 2014, pp. 1-4.

[7] T. B. Kumar, M. Kaixue, Y. Kiat Seng, and L. Wei Meng, "A 12-GHz high output power amplifier using 0.18 μm SiGe BiCMOS for low power applications," in *IEEE Asia Pacific Conference on Circuits and Systems (APCCAS)*, 2012, pp. 180-183.

[8] N. Mahalingam, M. Kaixue, Y. Kiat Seng, and L. Wei Meng, "Coupled dual LC tanks based ILFD with low injection power and compact size," *IEEE Microw. Wireless Compon. Lett.,* vol. 24, pp. 105-107, 2014.

[9] N. Mahalingam, M. Kaixue, Y. Kiat Seng, and L. Wei Meng, "*K*-band high-PAE wide-tuning-range VCO using triple-coupled *LC* tanks," *IEEE Trans. Circuits Syst. II, Exp. Briefs,* vol. 60, pp. 736-740, 2013.

[10] K. Ma, S. Mou, and K. S. Yeo, "Miniaturized 60-GHz on-chip multimode quasi-elliptical bandpass filter," *IEEE Electron Device Letters,* vol. 34, pp. 945-947, 2013.

[11] K. Ma, S. Mou, and K. S. Yeo, "A miniaturized millimeter-wave standing-wave filtering switch with high P1dB," *IEEE Trans. Microw. Theory Tech.,* vol. 61, pp. 1505-1515, 2013.

[12] F. Meng, K. Ma, K. S. Yeo, S. Xu, C. C. Boon, and W. M. Lim, "Miniaturized 3-bit phase shifter for 60 GHz phased-array in 65 nm CMOS technology," *IEEE Microw. Wireless Compon. Lett.,* vol. 24, pp. 50-52, 2014.

[13] W. T. Li, Y. C. Chiang, J. H. Tsai, H. Y. Yang, J. H. Cheng, and T. W. Huang, "60-GHz 5-bit phase shifter with integrated VGA phase-error compensation," *IEEE Trans. Microw. Theory Tech.,* vol. 61, pp. 1224 - 1235, 2013.

[14] K. Ma, N. Yan, K. S. Yeo, and W. M. Lim, "Miniaturized 40-60 GHz on-chip balun with capacitive loading compensation," *IEEE Electron Device Letters,* vol. 35, pp. 434-436, 2014.

[15] K. Lingkai, Dongjin, S., and Alon, E., "A 50mW-TX 65mW-RX 60GHz 4-element phased-array transceiver with integrated antennas in 65nm CMOS," in *IEEE Int. Solid-State Circuits Conf. (ISSCC) Dig.*, 2013, pp. 234-235.

[16] V. Vidojkovic, *et al.*, "A low-power radio chipset in 40nm LP CMOS with beamforming for 60GHz high-data-rate wireless communication," in *IEEE Int. Solid-State Circuits Conf. (ISSCC) Dig.*, 2013, pp. 236-237.

[17] J.-L. Kuo, *et al.*, "60-GHz four-element phased-array transmit/receive system-in-package using phase compensation techniques in 65-nm flip-chip CMOS process," *IEEE Trans. Microw. Theory Tech.,* vol. 60, pp. 743-756, 2012.

[18] W. Shin, B. H. Ku, O. Inac, Y. C. Ou, and G. M. Rebeiz, "A 108-114 GHz 4x4 wafer-scale phased array transmitter with high-efficiency on-chip antennas," *IEEE J. Solid-State Circuits,* vol. PP, pp. 1-15, 2013.

[19] F. Golcuk, Kanar, T., and Rebeiz, G. M., "A 90-100-GHz 4 x 4 SiGe BiCMOS polarimetric transmit/receive phased array with simultaneous receive-beams capabilities," *IEEE Trans. Microw. Theory Tech.,* vol. 61, pp. 3099-3114, 2013.

[20] S. Donghyup, K. Choul-Young, K. Dong-Woo, and G. M. Rebeiz, "A high-power packaged four-element X-band phased-array transmitter in 0.13-μm CMOS for radar and communication systems," *IEEE Trans. Microw. Theory Tech.,* vol. 61, pp. 3060-3071, 2013.

[21] M. Elkhouly, S. Glisic, C. Meliani, F. Ellinger, and J. C. Scheytt, "220 - 250-GHz phased-array circuits in 0.13-μm SiGe BiCMOS technology," *IEEE Trans. Microw. Theory Tech.,* vol. 61, pp. 3115-3127, 2013.

[22] S. Shahramian, Y. Baeyens, N. Kaneda, and Y. K. Chen, "A 70-100 GHz direct-conversion transmitter and receiver phased array chipset demonstrating 10 Gb/s wireless link," *IEEE J. Solid-State Circuits,* vol. 48, pp. 1113-1125, 2013.

[23] A. Natarajan, B. Floyd, and A. Hajimiri, "A bidirectional RF-combining 60GHz phased-array front-end," in *IEEE Int. Solid-State Circuits Conf. (ISSCC) Dig.*, 2007, pp. 202-597.

[24] E. Cohen, C. Jakobson, S. Ravid, and D. Ritter, "A thirty two element phased-array transceiver at 60GHz with RF-IF conversion block in 90nm flip chip CMOS process," in *IEEE Radio Frequency Integrated Circuits Symp. (RFIC) Dig.*, 2010, pp. 457-460.

[25] W. L. Chan, J. R. Long, M. Spirito, and J. J. Pekarik, "A 60GHz-band 2 x 2 phased-array transmitter in 65nm CMOS," in *IEEE Int. Solid-State Circuits Conf. (ISSCC) Dig.*, 2010, pp. 42-43.

[26] W. Cheng, *et al.*, "0.34-THz wireless link based on high-order modulation for future wireless local area network applications," *IEEE Trans. THz Science Technol.,* vol. 4, pp. 75-85, 2014.

[27] C. Gang, P. Jie, Y. Fei, Z. Xiao Yang, Z. L. Sun, and T. J. Cui, "Terahertz-wave imaging system based on backward wave oscillator," *IEEE Trans. THz Sci. Technol.,* vol. 2, pp. 504-512, 2012.

[28] F. Friederich, *et al.*, "THz active imaging systems with real-time capabilities," *IEEE Trans. THz Sci. Technol.,* vol. 1, pp. 183-200, 2011.

[29] Z. Chen, C.-C. Wang, H.-C. Yao, and P. Heydari, "A BiCMOS W-band 2×2 focal-plane array with on-chip antenna," *IEEE J. Solid-State Circuits,* vol. 47, pp. 2355-2371, 2012.

[30] H. Kazemi, *et al.*, "Low cost modular integrated horn antenna array using heterojunction barrier diode detectors," in *IEEE MTT-S Int. Microw. Symp. Dig.*, 2008, pp. 297-300.

[31] D. J. Frank, "Power-constrained CMOS scaling limits," *IBM Journal of Research and Development,* vol. 46, pp. 235-244, 2002.

[32] A. Ghaffari, E. A. M. Klumperink, F. van Vliet, and B. Nauta, "A 4-element phased-array system with simultaneous spatial- and frequency-domain filtering at the antenna inputs," *IEEE J. Solid-State Circuits,* vol. 49, pp. 1303-1316, 2014.

[33] C. W. Byeon, C. H. Yoon, and C. S. Park, "A 67-mW 10.7-Gb/s 60-GHz OOK CMOS transceiver for short-range wireless communications," *IEEE Trans. Microw. Theory Tech.,* vol. 61, pp. 3391-3401, 2013.

[34] E. Shumakher, J. Elkind, and D. Elad, "Key components of a 130 GHz dicke-radiometer SiGe RFIC," in *IEEE Silicon Monolith. Integr. Circuits in RF Syst. (SiRF)*, 2013, pp. 255-257.

[35] M. Uzunkol and G. M. Rebeiz, "A low-loss 50-70 GHz SPDT switch in 90 nm CMOS," *IEEE J. Solid-State Circuits,* vol. 45, pp. 2003-2007, 2010.

[36] R. L. Schmid, P. Song, C. T. Coen, A. C. Ulusoy, and J. D. Cressler, "On the analysis and design of low-loss single-pole double-throw W-band switches utilizing saturated SiGe HBTs," *IEEE Trans. Microw. Theory Tech.,* vol. 62, pp. 2755-2767, 2014.

[37] A. C. Ulusoy, *et al.*, "A low-loss and high isolation D-band SPDT switch utilizing deep-saturated SiGe HBTs," *IEEE Microw. Wireless Compon. Lett.,* vol. 24, pp. 400-402, 2014.

[38] M. Uzunkol and G. M. Rebeiz, "140-220 GHz SPST and SPDT switches in 45 nm CMOS SOI," *IEEE Microw. Wireless Compon. Lett.,* vol. 22, pp. 412-414, 2012.

[39] E. Laskin, *et al.*, "Nanoscale CMOS Transceiver Design in the 90-170-GHz Range," *IEEE Trans. Microw. Theory Tech.,* vol. 57, pp. 3477-3490, 2009.

[40] A. Tomkins, P. Garcia, and S. P. Voinigescu, "A Passive W-Band Imaging Receiver in 65-nm Bulk CMOS," *IEEE J. Solid-State Circuits,* vol. 45, pp. 1981-1991, 2010.

[41] E. Adabi and A. M. Niknejad, "A mm-wave transformer based transmit/receive switch in 90nm CMOS technology," in *European Microw. Conf. (EuMC)*, 2009, pp. 389-392.

[42] F. Meng, K. Ma, and K. S. Yeo, "A 130-to-180GHz 0.0035mm^2 SPDT switch with 3.3dB loss and 23.7dB isolation in 65nm bulk CMOS," in *IEEE Int. Solid-State Circuits Conf. (ISSCC) Dig.*, 2015, pp. 34-36.

[43] F. Meng, K. Ma, K. S. Yeo, C. C. Boon, W. M. Lim, and S. Xu, "A 220-285 GHz SPDT switch in 65-nm CMOS using switchable resonator concept," *IEEE Trans. THz Sci. Technol.,* vol. 5, pp. 649-651, Jul. 2015.

[44] J. Mayer, "MOS models and circuit simulation," *RVA Review,* vol. 32, pp. 42-63, 1971.

[45] C. C. Enz and C. Yuhua, "MOS transistor modeling for RF IC design," *IEEE J. Solid-State Circuits,* vol. 35, pp. 186-201, 2000.

[46] D. M. Pozar, *Microwave Engineering, 4th ed*: New York: Wiley, 2012.

[47] U. Decanis, A. Ghilioni, E. Monaco, A. Mazzanti, and F. Svelto, "A low-noise quadrature VCO based on magnetically coupled resonators and a wideband frequency divider at millimeter waves," *IEEE J. Solid-State Circuits,* vol. 46, pp. 2943-2955, 2011.

[48] L. Wen-Chian and C. Huey-Ru, "A 40-110 GHz high-isolation CMOS traveling-wave T/R switch by using parallel inductor," in *IEEE MTT-S Int. Microw. Symp. Dig.*, 2015, pp. 1-3.

[49] J. W. May and G. M. Rebeiz, "Design and characterization of *W*-band SiGe RFICs for passive millimeter-wave imaging," *IEEE Trans. Microw. Theory Tech.,* vol. 58, pp. 1420-1430, 2010.

[50] Q. Liu, H. Liu, and Y. Liu, "Compact ultra-wideband 90° phase shifter using short-circuited stub and weak coupled line," *Electronics Letters,* vol. 50, pp. 1454-1456, 2014.

[51] H. Lee and B. Min, "W-band CMOS 4-Bit phase shifter for high power and phase compression points," *IEEE Trans. Circuits Syst. II, Exp. Briefs,* vol. PP, pp. 1-1, 2014.

[52] P.-S. Huang and H.-C. Lu, "Broadband differential phase-shifter design using bridged T-type bandpass network," *IEEE Trans. Microw. Theory Tech.,* vol. 62, pp. 1470-1479, 2014.

[53] X. Tang and K. Mouthaan, "Large bandwidth digital phase shifters with all-pass, high-pass, and low-pass networks," *IEEE Trans. Microw. Theory Tech.,* vol. 61, pp. 2325-2331, 2013.

[54] I. Kalyoncu, E. Ozeren, M. Kaynak, and Y. Gurbuz, "A 4-bit SiGe passive phase shifter for X-band phased arrays," in *IEEE Silicon Monolith. Integr. Circuits in RF Syst. (SiRF)*, 2013, pp. 210-212.

[55] E. Topak, J. Hasch, C. Wagner, and T. Zwick, "A novel millimeter-wave dual-fed phased array for beam steering," *IEEE Trans. Microw. Theory Tech.,* vol. 61, pp. 3140-3147, 2013.

[56] S. Sanghoon, L. Jeon, and K. Jeong-Geun, "A compact X-band bi-directional phased-array T/R chipset in 0.13 μm CMOS technology," *IEEE Trans. Microw. Theory Tech.,* vol. 61, pp. 562-569, 2013.
[57] L. G. Sodin, "Method of synthesizing a beam-forming device for the N-beam and N-element array antenna, for any N," *IEEE Trans. Antennas Propag.,* vol. 60, pp. 1771-1776, 2012.
[58] M. Sorn, R. Lech, and J. Mazur, "Simulation and experiment of a compact wideband 90 differential phase shifter," *IEEE Trans. Microw. Theory Tech.,* vol. 60, pp. 494-501, 2012.
[59] X. Tang and K. Mouthaan, "Design of large bandwidth phase shifters using common mode all-pass networks," *IEEE Microw. Wireless Compon. Lett.,* vol. 22, pp. 55-57, 2012.
[60] Y.-Y. Huang, H. Jeon, Y. Yoon, W. Woo, C.-H. Lee, and J. S. Kenney, "An ultra-compact, linearly-controlled variable phase shifter designed with a novel RC poly-phase filter," *IEEE Trans. Microw. Theory Tech.,* vol. 60, pp. 301-310, 2012.
[61] A. M. Abbosh, "Compact tunable reflection phase shifters using short section of coupled lines," *IEEE Trans. Microw. Theory Tech.,* vol. PP, pp. 1-1, 2012.
[62] A. Asoodeh and M. Atarodi, "A full 360° vector-sum phase shifter with very low RMS phase error over a wide bandwidth " *IEEE Trans. Microw. Theory Tech.,* vol. 60, pp. 1626-1634, 2012.
[63] Y. Park, "A CMOS voltage controlled continuous phase shifter with active loss compensation," *IEEE Microw. Wireless Compon. Lett.,* vol. PP, pp. 1-1, 2012.
[64] Y.-C. Chiang, W.-T. Li, J.-H. Tsai, and T.-W. Huang, "A 60GHz digitally controlled 4-bit phase shifter with 6-ps group delay deviation," in *IEEE MTT-S Int. Microw. Symp. Dig.*, 2012, pp. 1-3.
[65] S. Y. Kim and G. M. Rebeiz, "A low-power BiCMOS 4-element phased array receiver for 76-84 GHz radars and communication systems," *IEEE J. Solid-State Circuits,* vol. 47, pp. 359-367, 2012.
[66] W. Shin and G. M. Rebeiz, "60 GHz active phase shifter using an optimized quadrature all-pass network in 45nm CMOS," in *IEEE MTT-S Int. Microw. Symp. Dig.*, 2012, pp. 1-3.
[67] M. Uzunkol and G. M. Rebeiz, "A 65 GHz LNA/phase shifter with 4.3 dB NF using 45 nm CMOS SOI," *IEEE Microw. Wireless Compon. Lett.,* vol. 22, pp. 530-532, 2012.
[68] A. B. Nguyen and J.-W. Lee, "A K-Band CMOS Phase Shifter MMIC Based on a Tunable Composite Metamaterial," *IEEE Microw. Wireless Compon. Lett.,* vol. 21, pp. 311-313, 2011.
[69] M. Tabesh, A. Arbabian, and A. Niknejad, "60GHz low-loss compact phase shifters using a transformer-based hybrid in 65nm CMOS," in *IEEE Custom Integrated Circuits Conf.*, 2011, pp. 1-4.
[70] S. Emami, *et al.*, "A 60GHz CMOS phased-array transceiver pair for multi-Gb/s wireless communications," in *IEEE Int. Solid-State Circuits Conf. (ISSCC) Dig.*, 2011, pp. 164-166.

[71] C.-W. Wang, H.-S. Wu, and C. K. C. Tzuang, "CMOS passive phase shifter with group-delay deviation of 6.3 ps at K-band," *IEEE Trans. Microw. Theory Tech.,* vol. 59, pp. 1778-1786, 2011.

[72] N. Somjit, G. Stemme, and J. Oberhammer, "Power handling analysis of high-power W-band all-silicon MEMS phase shifters," *IEEE Trans. Electron Devices,* vol. 58, pp. 1548-1555, 2011.

[73] S. Gong, H. Shen, and N. S. Barker, "A 60-GHz 2-bit switched-line phase shifter using SP4T RF-MEMS switches," *IEEE Trans. Microw. Theory Tech.,* vol. 59, pp. 894-900, 2011.

[74] K. Raczkowski, W. De Raedt, B. Nauwelaers, and P. Wambacq, "A wideband beamformer for a phased-array 60GHz receiver in 40nm digital CMOS," in *IEEE Int. Solid-State Circuits Conf. (ISSCC) Dig.*, 2010, pp. 40-41.

[75] M. Fakharzadeh, M. R. Nezhad-Ahmadi, B. Biglarbegian, J. Ahmadi-Shokouh, and S. Safavi-Naeini, "CMOS phased array transceiver technology for 60 GHz wireless applications," *IEEE Trans. Antennas Propag.,* vol. 58, pp. 1093-1104, 2010.

[76] D.-W. Kang, K.-J. Koh, and G. M. Rebeiz, "A Ku-band two-antenna four-simultaneous beams SiGe BiCMOS phased array receiver," *IEEE Trans. Microw. Theory Tech.,* vol. 58, pp. 771-780, 2010.

[77] M. Hangai, M. Hieda, N. Yunoue, Y. Sasaki, and M. Miyazaki, "S- and C-band ultra-compact phase shifters based on all-pass networks," *IEEE Trans. Microw. Theory Tech.,* vol. 58, pp. 41-47, 2010.

[78] K.-J. Koh and G. M. Rebeiz, "A 6-18 GHz 5-bit active phase shifter," in *IEEE MTT-S Int. Microw. Symp. Dig.*, 2010, pp. 792-795.

[79] M. Meghdadi, M. Azizi, M. Kiani, A. Medi, and M. Atarodi, "A 6-Bit CMOS phase shifter for S-band," *IEEE Trans. Microw. Theory Tech.,* vol. 58, pp. 3519-3526, 2010.

[80] C. W. Wang, H. S. Wu, and C. K. C. Tzuang, "A miniaturized power combiner for compact design of CMOS phase shifter at K-Band," in *IEEE MTT-S Int. Microw. Symp. Dig.*, 2010, pp. 121-124.

[81] E. Cohen, C. G. Jakobson, S. Ravid, and D. Ritter, "A bidirectional TX/RX four-element phased array at 60 GHz with RF-IF conversion block in 90-nm CMOS process," *IEEE Trans. Microw. Theory Tech.,* vol. 58, pp. 1438-1446, 2010.

[82] J. C. Rock, *et al.*, "Design, fabrication and evaluation of a MEMS-based, Ka-band, 16-element sub-array," in *IEEE Aerospace Conf.*, 2010, pp. 1-17.

[83] P. Perez-Lara, I. Molina-Fernandez, J. G. Wanguemert-Perez, and A. Rueda-Perez, "Broadband five-port direct receiver based on low-pass and high-pass phase shifters," *IEEE Trans. Microw. Theory Tech.,* vol. 58, pp. 849-853, 2010.

[84] Y. Yu, P. G. M. Baltus, G. Anton de, E. van der Heijden, C. S. Vaucher, and A. H. M. Van Roermund, "A 60 GHz phase shifter integrated with LNA and PA in 65 nm CMOS for phased array systems," *IEEE J. Solid-State Circuits,* vol. 45, pp. 1697-1709, 2010.

[85] W. Aerts, P. Delmotte, and G. A. E. Vandenbosch, "Conceptual study of analog baseband beam forming: Design and measurement of an eight-by-eight phased array," *IEEE Trans. Antennas Propag.,* vol. 57, pp. 1667-1672, 2009.

[86] B. Biglarbegian, M. R. Nezhad-Ahmadi, M. Fakharzadeh, and S. Safavi-Naeini, "Millimeter-wave reflective-type phase shifter in CMOS technology," *IEEE Microw. Wireless Compon. Lett.,* vol. 19, pp. 560-562, 2009.

[87] M.-D. Tsai and A. Natarajan, "60GHz passive and active RF-path phase shifters in silicon," in *IEEE RFIC Symp.Dig.*, 2009, pp. 223-226.

[88] V. Puyal, D. Dragomirescu, C. Villeneuve, R. Jinyu, P. Pons, and R. Plana, "Frequency scalable model for MEMS capacitive shunt switches at millimeter-wave frequencies," *IEEE Trans. Microw. Theory Tech.,* vol. 57, pp. 2824-2833, 2009.

[89] I. Sarkas, M. Khanpour, A. Tomkins, P. Chevalier, P. Garcia, and S. P. Voinigescu, "W-band 65-nm CMOS and SiGe BiCMOS transmitter and receiver with lumped I-Q phase shifters," in *IEEE Radio Frequency Integrated Circuits Symp. (RFIC)*, 2009, pp. 441-444.

[90] S. Y. Kim and G. M. Rebeiz, "A 4-Bit Passive phase shifter for automotive radar applications in 0.13 μm CMOS," in *IEEE Compound Semiconductor Integrated Circuit Symp. (CSIC) Dig.*, 2009, pp. 1-4.

[91] K.-J. Koh, J. W. May, and G. M. Rebeiz, "A millimeter-wave (40-45 GHz) 16-element phased-array transmitter in 0.18-μm SiGe BiCMOS technology," *IEEE J. Solid-State Circuits,* vol. 44, pp. 1498-1509, 2009.

[92] D.-W. Kang, J.-G. Kim, B.-W. Min, and G. M. Rebeiz, "Single and four-element Ka-band transmit/receive phased-array Silicon RFICs with 5-bit amplitude and phase control," *IEEE Trans. Microw. Theory Tech.,* vol. 57, pp. 3534-3543, 2009.

[93] C.-S. Lin, S.-F. Chang, and W.-C. Hsiao, "A full-360 reflection-type phase shifter with constant insertion loss," *IEEE Microw. Wireless Compon. Lett.,* vol. 18, pp. 106-108, 2008.

[94] J.-C. Wu, T.-Y. Chin, S.-F. Chang, and C.-C. Chang, "2.45-GHz CMOS reflection-type phase-shifter MMICs with minimal loss variation over quadrants of phase-shift range," *IEEE Trans. Microw. Theory Tech.,* vol. 56, pp. 2180-2189, 2008.

[95] Y. Yu, *et al.*, "A 60GHz digitally controlled phase shifter in CMOS," in *European Solid-State Circuits Conf. (ESSCIRC) Dig.*, 2008, pp. 250-253.

[96] M. A. Morton and J. Papapolymerou, "A packaged MEMS-based 5-bit X-band high-pass/low-pass phase shifter," *IEEE Trans. Microw. Theory Tech.,* vol. 56, pp. 2025-2031, 2008.

[97] I. J. Bahl and D. Conway, "L- and S-band compact octave bandwidth 4-bit MMIC phase shifters," *IEEE Trans. Microw. Theory Tech.,* vol. 56, pp. 293-299, 2008.

[98] B.-W. Min and G. M. Rebeiz, "Single-ended and differential Ka-band BiCMOS phased array front-ends," *IEEE J. Solid-State Circuits,* vol. 43, pp. 2239-2250, 2008.

[99] D. Betancourt and C. del Rio Bocio, "A novel methodology to feed phased array antennas," *IEEE Trans. Antennas Propag.*, vol. 55, pp. 2489-2494, 2007.

[100] C.-S. Lin, S.-F. Chang, C.-C. Chang, and Y.-H. Shu, "Design of a reflection-type phase shifter with wide relative phase shift and constant insertion loss," *IEEE Trans. Microw. Theory Tech.,* vol. 55, pp. 1862-1868, 2007.

[101] D.-W. Kang and S. Hong, "A 4-bit CMOS phase shifter using distributed active switches," *IEEE Trans. Microw. Theory Tech.,* vol. 55, pp. 1476-1483, 2007.

[102] H. Zarei, C. T. Charles, and D. J. Allstot, "Reflective-type phase shifters for multiple-antenna transceivers," *IEEE Trans. Circuits Syst. I, Reg. Papers,* vol. 54, pp. 1647-1656, 2007.

[103] S. E. Shih, *et al.*, "A W-band 4-Bit phase shifter in multilayer scalable array systems," in *IEEE Compound Semiconductor Integrated Circuit Symp. (CSIC) Dig.*, 2007, pp. 1-4.

[104] L. Y. V. Chen, R. Forse, A. H. Cardona, T. C. Watson, and R. York, "Compact analog phase shifters using thin-film (Ba,Sr)TiO3 varactors," in *IEEE MTT-S Int. Microw. Symp. Dig.*, 2007, pp. 667-670.

[105] K.-J. Koh and G. M. Rebeiz, "0.13-µm CMOS phase shifters for X-, Ku-, and K-band phased arrays," *IEEE J. Solid-State Circuits,* vol. 42, pp. 2535-2546, 2007.

[106] A. Natarajan, A. Komijani, X. Guan, A. Babakhani, and A. Hajimiri, "A 77-GHz phased-array transceiver with on-chip antennas in Silicon: Transmitter and local LO-path phase shifting," *IEEE J. Solid-State Circuits,* vol. 41, pp. 2807-2819, 2006.

[107] D.-W. Kang, H.-D. Lee, C.-H. Kim, and S. Hong, "Ku-band MMIC phase shifter using a parallel resonator with 0.18-µm CMOS technology," *IEEE Trans. Microw. Theory Tech.,* vol. 54, pp. 294-301, 2006.

[108] N. Kingsley and J. Papapolymerou, "Organic "wafer-scale" packaged miniature 4-bit RF MEMS phase shifter," *IEEE Trans. Microw. Theory Tech.,* vol. 54, pp. 1229-1236, 2006.

[109] M. A. Morton, J. P. Comeau, J. D. Cressler, M. Mitchell, and J. Papapolymerou, "Sources of phase error and design considerations for silicon-based monolithic high-pass/low-pass microwave phase shifters," *IEEE Trans. Microw. Theory Tech.,* vol. 54, pp. 4032-4040, 2006.

[110] D.-W. Kang, H. D. Lee, C.-H. Kim, and S. Hong, "Ku-band MMIC phase shifter using a parallel resonator with 0.18-/spl mu/m CMOS technology," *IEEE Trans. Microw. Theory Tech.,* vol. 54, pp. 294-301, 2006.

[111] Y.-X. Guo, Z.-Y. Zhang, and L.-C. Ong, "Improved wide-band Schiffman phase shifter," *IEEE Trans. Microw. Theory Tech.,* vol. 54, pp. 1196-1200, 2006.

[112] T. M. Hancock and G. M. Rebeiz, "A 12-GHz SiGe phase shifter with integrated LNA," *IEEE Trans. Microw. Theory Tech.,* vol. 53, pp. 977-983, 2005.

[113] T. Yamaji, D. Kurose, O. Watanabe, S. Obayashi, and T. Itakura, "A four-input beam-forming downconverter for adaptive antennas," *IEEE J. Solid-State Circuits,* vol. 38, pp. 1619-1625, 2003.
[114] F. Ellinger, R. Vogt, and W. Bachtold, "Ultracompact reflective-type phase shifter MMIC at C-band with 360° phase-control range for smart antenna combining," *IEEE J. Solid-State Circuits,* vol. 37, pp. 481-486, 2002.
[115] M. Teshiba, R. Van Leeuwen, G. Sakamoto, and T. Cisco, "A SiGe MMIC 6-bit PIN diode phase shifter," *IEEE Microw. Wireless Compon. Lett.,* vol. 12, pp. 500-501, 2002.
[116] N. C. Karmakar and M. E. Bialkowski, "A beam-forming network for a circular switched-beam phased array antenna," *IEEE Microw. Wireless Compon. Lett.,* vol. 11, pp. 7-9, 2001.
[117] F. Ellinger, R. Vogt, and W. Bachtold, "Compact reflective-type phase-shifter MMIC for C-band using a lumped-element coupler," *IEEE Trans. Microw. Theory Tech.,* vol. 49, pp. 913-917, 2001.
[118] K. Miyaguchi*, et al.*, "An ultra-broad-band reflection-type phase-shifter MMIC with series and parallel LC circuits," *IEEE Trans. Microw. Theory Tech.,* vol. 49, pp. 2446-2452, 2001.
[119] F. Ellinger, U. Lott, and W. Bachtold, "An antenna diversity MMIC vector modulator for HIPERLAN with low power consumption and calibration capability," *IEEE Trans. Microw. Theory Tech.,* vol. 49, pp. 964-969, 2001.
[120] K. Maruhashi, H. Mizutani, and K. Ohata, "Design and performance of a Ka-band monolithic phase shifter utilizing nonresonant FET switches," *IEEE Trans. Microw. Theory Tech.,* vol. 48, pp. 1313-1317, 2000.
[121] C. F. Campbell and S. A. Brown, "A compact 5-bit phase shifter MMIC for K-band satellite communication systems," in *IEEE MTT-S Int. Microw. Symp. Dig.*, 2000, pp. 217-220.
[122] M. Chua and K. W. Martin, "1 GHz programmable analog phase shifter for adaptive antennas," in *IEEE Custom Integrated Circuits Conf.*, 1998, pp. 71-74.
[123] T. Ohira, Y. Suzuki, H. Ogawa, and K. Kamitsuna, "Megalithic microwave signal processing for phased-array beamforming and steering," *IEEE Trans. Microw. Theory Tech.,* vol. 45, pp. 2324-2332, 1997.
[124] M. Mahfoudi and J. I. Alonso, "A simple technique for the design of MMIC 90° phase-difference networks," *IEEE Trans. Microw. Theory Tech.,* vol. 44, pp. 1694-1702, 1996.
[125] S. Lucyszyn and I. D. Robertson, "Synthesis techniques for high performance octave bandwidth 180° analog phase shifters," *IEEE Trans. Microw. Theory Tech.,* vol. 40, pp. 731-740, 1992.
[126] D. Adler and R. Popovich, "Broadband switched-bit phase shifter using all-pass networks," in *IEEE MTT-S Int. Microw. Symp. Dig.*, 1991, pp. 265-268 vol.1.
[127] J. L. R. Quirarte and J. P. Starski, "Synthesis of Schiffman phase shifters," *IEEE Trans. Microw. Theory Tech.,* vol. 39, pp. 1885-1889, 1991.

[128] C. Moye, G. Sakamoto, and M. Brand, "A compact broadband, six-bit MMIC phasor with integrated digital drivers," in *IEEE Microw. and Millimeter-wave Monolithic Circuit Symp.*, 1990, pp. 123-126.

[129] M. J. Schindler and M. E. Miller, "A 3 bit K/Ka band MMIC phase shifter," in *IEEE Microw. and Millimeter-wave Monolithic Circuit Symp.*, 1988, pp. 95-98.

[130] D. M. Krafcsik, S. A. Imhoff, D. E. Dawson, and A. L. Conti, "A dual-varactor analog phase shifter operating at 6 to 18 GHz," *IEEE Trans. Microw. Theory Tech.,* vol. 36, pp. 1938-1941, 1988.

[131] A. W. Jacomb-Hood, D. Seielstad, and J. D. Merrill, "A three-bit monolithic phase shifter at V-band," in *IEEE Microw. and Millimeter-wave Monolithic Circuit Symp.*, 1987, pp. 81-84.

[132] H. A. Atwater, "Circuit design of the loaded-line phase shifter," *IEEE Trans. Microw. Theory Tech.,* vol. 33, pp. 626-634, 1985.

[133] C. Andricos, I. J. Bahl, and E. L. Griffin, "C-band 6-bit GaAs monolithic phase shifter," *IEEE Trans. Electron Devices,* vol. 32, pp. 2760-2765, 1985.

[134] Y. Ayasli, S. W. Miller, R. L. Mozzi, and L. K. Hanes, "Wideband S-C band monolithic phase shifter," in *IEEE Microw. and Millimeter-wave Monolithic Circuit Symp.*, 1984, pp. 11-13.

[135] I. J. Bahl and K. C. Gupta, "Design of loaded-line p-i-n diode phase shifter circuits," *IEEE Trans. Microw. Theory Tech.,* vol. 28, pp. 219-224, 1980.

[136] G. A. Swartz, A. Rosen, P. T. Ho, and A. Schwarzmann, "Low-loss p-i-n diode for high-power MIC phase shifter," *IEEE Trans. Electron Devices,* vol. 25, pp. 1297-1301, 1978.

[137] W. A. Davis, "Design equations and bandwidth of loaded-line phase shifters (short papers)," *IEEE Trans. Microw. Theory Tech.,* vol. 22, pp. 561-563, 1974.

[138] T. Yahara, "A note on designing digital diode-loaded-line phase shifters (short papers)," *IEEE Trans. Microw. Theory Tech.,* vol. 20, pp. 703-704, 1972.

[139] R. V. Garver, "Broad-band diode phase shifters," *IEEE Trans. Microw. Theory Tech.,* vol. 20, pp. 314-323, 1972.

[140] J. F. White, "Review of semiconductor microwave phase shifters," *Proceedings of the IEEE,* vol. 56, pp. 1924-1931, 1968.

[141] J. F. White, "High power, p-i-n diode controlled, microwave transmission phase shifters," *IEEE Trans. Microw. Theory Tech.,* vol. 13, pp. 233-242, 1965.

[142] B. M. Schiffman, "A new class of broad-band microwave 90-degree phase shifters," *IRE Trans. Microw. Theory Tech.,* vol. 6, pp. 232-237, 1958.

[143] L. Kuang, B. Chi, H. Jia, Z. Ye, W. Jia, and Z. Wang, "Co-design of 60-GHz wideband front-end IC with on-chip T/R switch based on passive macro-modeling," *IEEE Trans. Microw. Theory Tech.,* vol. 62, pp. 2743-2754, 2014.

[144] H. Jin, X. Yong-Zhong, and Z. Yue Ping, "Analysis and design of 60-GHz SPDT switch in 130-nm CMOS," *IEEE Trans. Microw. Theory Tech.,* vol. 60, pp. 3113-3119, 2012.

[145] M.-H. Tsai, S. S. H. Hsu, F.-L. Hsueh, C.-P. Jou, and T. Yeh, "Design of 60-GHz low-noise amplifiers with low NF and robust ESD protection in 65-nm CMOS," *IEEE Trans. Microw. Theory Tech.,* vol. 61, pp. 553-561, 2013.

[146] W.-T. Li, *et al.*, "Parasitic-insensitive linearization methods for 60-GHz 90-nm CMOS LNAs," *IEEE Trans. Microw. Theory Tech.,* vol. 60, pp. 2512-2523, 2012.
[147] P. M. Farahabadi and K. Moez, "A dual-mode highly efficient 60 GHz power amplifier in 65 nm CMOS," in *IEEE Radio Frequency Integrated Circuits Symp. (RFIC) Dig.*, 2014, pp. 155-158.
[148] F. Wei, Y. Hao, S. Yang, and Y. Kiat Seng, "A 2-D distributed power combining by metamaterial-based zero phase shifter for 60-GHz power amplifier in 65-nm CMOS," *IEEE Trans. Microw. Theory Tech.,* vol. 61, pp. 505-516, 2013.
[149] S. Shiba, *et al.*, "F-band bidirectional amplifier using 75-nm InP HEMTs," in *IEEE Compound Semicond. Integr. Circuit Symp. (CSICS)*, 2012, pp. 1-4.

More
Books!
yes

Made in the USA
Columbia, SC
04 March 2025

54678438R00067